Migrants and Memory

Migrants and Memory
The Forgotten "Postcolonials"

Edited by

Mícheál Ó hAodha

CAMBRIDGE SCHOLARS PUBLISHING

Migrants and Memory: The Forgotten "Postcolonials", edited by Mícheál Ó hAodha

This book first published 2007 by

Cambridge Scholars Publishing

15 Angerton Gardens, Newcastle, NE5 2JA, UK

British Library Cataloguing in Publication Data
A catalogue record for this book is available from the British Library

ISBN 1-84718-344-1; ISBN 13: 9781847183446

TABLE OF CONTENTS

INTRODUCTION

This volume is an exploration of the image that is the Traveller/Gypsy, the nomad, the migrant and the outsider/"Other" within the frame of articulation that is European representational culture. It is another small addition to the burgeoning subject areas that are Migration/Diaspora Studies and Irish Studies. The global flows of mass-migration and mass-media dissemination that are a commonplace today have overcome centuries-old physical divides as evidenced between diverse peoples and cultural groups. They have also created innovative spaces for the exploration of issues relating to cross-cultural and identity representation. Academics, cultural analysts, writers and artists all increasingly work in transnational and transcultural spaces, spaces which Arjun Appadurai (2001)has referred to as "imagined worlds" (2001, 329). These new creative spaces are environments where alliances and allegiances coalesce, dissolve, and coalesce again as relating to particular ideas and images which are continuously re-staged across, rather than within, stable nationalist cultural narratives. Postcolonial diasporas have served to both intensify and accelerate this undermining and reconfiguring of (hitherto) dominant cultural narratives. Nowhere is this disruption of narrative more in evidence than in the case of diaspora peoples and cultural groups such as Travellers, Roma (Gypsies) and other migrant groups. As outlined in the essays that comprise this book the diverse and often-reductionist representations of such "outsider" groups disrupt the narratives which define dominant cultures and hybridize the discourse. In the process they serve to remind us that cultural identities are always fluid and heterogeneous and may disrupt such "traditional" concepts as the "nation" or what we may define as "home" or "abroad". Volumes such as the following hope to provide a forum for an enhanced dialogue between the old and the new, the local and the foreign, the indigene and the migrant. Through such a dialogic process we can ensure that the voices "from the margin" can begin to be heard both inside and outside of the dominant discourse.

Mícheál Ó hAodha
University of Limerick, Ireland

CHAPTER ONE

THE CONSTRUCTION OF IRISH TRAVELLERS (AND GYPSIES) AS A 'PROBLEM'

ANTHONY DRUMMOND

From March through June 2005 I researched Irish Travellers' perceptions of, and experiences with, the criminal justice system in both Northern Ireland and the Republic of Ireland.[1] This essay draws on an extensive literature review that placed my research in context. As will become clear, across the island of Ireland lack of accommodation provision (especially transient sites)[2] by state parties is a key issue illuminating the situation of Irish Travellers as relating to criminal justice. Therefore, in the first instance, an account of the failure to adequately provide accommodation for Travellers in the Republic of Ireland is detailed, followed by an outline of similar failures in Northern Ireland. A summary of the social history and politics of each jurisdiction is then set down and analysed in order to consider official (and public) responses towards Irish Travellers. Thereafter, general societal perceptions (including those held by some police) of Travellers and Gypsies[3] as congenitally criminal are examined,

---

[1] As of early 2007 the title of my thesis, undertaken at the University of Ulster was: 'No Safe Space: Irish Travellers and Criminal Justice across the Island of Ireland'. During the research 25 Travellers, 9 members of criminal justice agencies, 6 non-Traveller support workers,1 priest and 1 member of a non-Traveller voluntary agency were interviewed.

[2] A transient site (sometimes referred to as a transit site) is a place that would legally enable Travellers to continue with their nomadic way of life and permit them to take up residence for weeks at a time.

[3] As will become evident, in the past, Irish Travellers have often been referred to by state parties as Gypsies (and itinerants/tinkers). Thus, in the past, the term of Gypsy has become synonymous with that of Travellers. However, Irish Travellers are distinct from other nomadic groups such as the Roma as they are indigenous to Ireland and the majority are Catholic and speak a language known as Cant, Shelta or Gammon.

followed by what is known on Travellers and the criminal justice system in Britain and lastly, existing literature on the situation of Irish Travellers and criminal justice across the island of Ireland is considered.

Broadly, the conceptual framework of my research concerned labelling theory, underpinned by the concept of sedentarism explained as, '...that system of ideas and practices which serves to normalise and reproduce sedentary modes of existence' whilst repressing nomadic modes of existence (McVeigh 1997: 9). The majority of articles emerging from the literature review also indicated that sedentarism includes the '...active and intentional incitement of fear and hatred of nomads', (McVeigh 1997: 9) obviously, including anti-Travellerism.[4]

The Background to Accommodation Provision for Travellers in the Republic of Ireland

As will become clear, the criminalisation of nomadic Travellers has forced many to forsake their nomadic traditions and become an urban as opposed to a rural and nationwide phenomenon. However, despite the movement of Travellers towards towns and metropolis, for a variety of reasons many remain marginalised from access to mainstream opportunities that the majority ethnic population take for granted such as relating to education and employment. In addition, concerns have been raised that increased contact with settled people has led to Travellers becoming involved in drug use and dealing. In fact it has been stated that, '...as we settle we have taken on the habits of the settled community, from their schools, on the street and in the clubs or bars' (Taylor 2004, cited in Fountain 2006: 61). Thus, the importance of charting the history of the failure to provide accommodation for nomadic Travellers and the subsequent increased and rapid urbanisation of them relative to a debate on criminal justice can not be underestimated.

Recently, Mulcahy and O'Mahony (2005: 36) observed that, '[m]uch of the background to difficulties between Travellers and the police can be traced back to the issue of [lack of] Traveller accommodation [per se]'. Indeed, the fact that provision of transient sites or static sites for Travellers was never a priority for the Irish Government can be traced back to the

---

[4] In order to comprehend the racism experienced by Travellers, the term anti-Travellerism is important. In itself the term implies that the racism suffered by Travellers is different to that experienced by other settled racial minority groups. The majority of major sociological works on racism have failed to consider the fact that Gypsies and Travellers are ethnic minorities and that they too experience racism.

1960's when the Irish State was keen that Irish Travellers become settled and reside in houses (see McDonagh 2001: 22).[5] The settlement of Irish Travellers was not the only intention of the Irish Government though. At the inaugural meeting of the Commission on Itinerancy in 1963, Parliamentary Secretary to the Minister for Justice, Charles Haughey, noted that '…there can be no final solution[6] to the problems created by itinerants until they are absorbed into the general community' (Commission on Itinerancy 1963:111; cited in EA 2006: 14).[7] As suggested by Court (1985: 1; cited in Mac Laughlin 1998: 424) Haughey's stance indicated that during the 1960's, the Irish were concerned to '…expunge Tinkerism[8] [from Irish society], not to vindicate or assess its significance'. Thus, as observed by Liègeois (1994: 199, cited in Gordon 1999: 291), '…whenever we look at how Gypsies and Travellers are treated' we are informed as to, '…the social history, [and] the politics……of those who are reacting to them'.

The fact that Haughey believed there could be 'no final solution' to the problems created by itinerants until they are absorbed into the general community' (CoI, 1963:111) serves as a socio–historical document and clearly indicates the official (political) stance vis-à-vis Irish Travellers during the 1960's. Haughey's (CoI 1963) remarks also demonstrated precisely how, '[t]he problems that Gypsies face [in this case, Irish Travellers] (of discrimination, lack of services and so on) are reconceptualised as Gypsies being the problem' (Gordon 1999: 279). Apposite to the debate concerning harsh new anti-trespass laws in 2002, outlined in due course, Gordon (1999: 279) also considered that:

> The creation of the 'Gypsy problem' not only vindicates government, institutions, and individuals that repeatedly discriminate and abuse Gypsies, but blames Gypsies for their predicament, thereby legitimising

---

[5] Irish Traveller McDonagh (2001: 22) is of the opinion that there is no such thing as a 'settled Traveller' stating that, 'Many Travellers live in houses today….but they still have nomadic leanings and would love to take to the road…. if only they could'.

[6] Haughey's use of the term 'final solution' mirrors Nazi ideology used against Jews, Gypsies, homosexuals and others such as those with disabilities, considered to be undesirables during The Holocaust (see Bauman 1989).

[7] For a detailed background on the treatment of Travellers from 1974 – 1998 across the Republic of Ireland see work by the following authors: also Gmelch, G. and Gmelch, S. (1974); Gmelch (1975); Gmelch G. and Gmelch S. (1976); Gmelch (1985); Gmelch, (1989); Helleiner, (1993); Molloy (1998).

[8] The term Tinker refers to Irish Travellers who were skilled in making pots and pans and buckets from tin.

> more discrimination and abuse, redefined as 'control' and 'punishment. Furthermore, just as had occurred within Nazi Germany, the construction of a race as a 'problem' [as insinuated by Haughey in the Republic of Ireland in 1963] legitimises measures to find a solution to the 'problem'.

The fact that 'itinerants/tinkers' (Travellers) were perceived of as a problem to be expunged was underlined in 1964 as '…the programme of Traveller settlement outlined in the Report of the Commission on Itinerancy [1963]…. [was]…adopted as government policy' (Helleiner, 1993: 187). Nevertheless, the Commission on Itinerancy acknowledged that, '…a housing shortage in urban areas would make immediate implementation difficult' (CI 1963: 54; in Helleiner 1993: 187) and consequently, some official sites were developed for Travellers.

The situation for Irish Travellers was largely ignored until 1983 when the Traveller People Review Body was formed. However, the Traveller People Review Body remained in favour of the provision of standard housing accommodation for Travellers (Molloy, 1998: 34). Nevertheless, a few years later, within Section 13 - (1) of the Housing Act 1988 the needs of Travellers described as those, '…persons belonging to the class of persons who traditionally pursue or have pursued a nomadic way of life..' were considered. In fact, in Section 13 – (2) it is stated that:

> A housing authority may provide, improve, manage and control sites for caravans used by persons to whom this section applies, and may carry out any works incidental to such provision, improvement, management or control, including the provision of services for such sites.

Despite the provisos outlined within the Housing Act 1988 above, adequate accommodation provision for Travellers was not met (NTACC 2004; Whyte 2002) and it continued to be provided on a piecemeal basis during the 1980's, when, for example, a segregated Traveller village was established in the city of Galway (Helleiner, 1993: 191).

There remained another silence on Travellers until the unremitting crisis for *all* of them led to the Irish Government setting up a Task Force on the Travelling Community in 1995 (see Molloy 1994: 34). In contrast to previous practices, Traveller support groups and Travellers themselves were widely consulted. The consultation signalled that an inclusive, more positive approach was being taken towards Travellers. The renewed approach towards Travellers recognised their '…unique culture and lifestyle' (Molloy 1998: 34) that should have been facilitated '…through the provision of a range of accommodation options, from group housing [schemes] to transit sites…' (*ibid:* 34). In fact, it was claimed that provision of transient sites would enable the nomadic lifestyle of

Travellers to, '…be accommodated and facilitated, in such a way as to enable them to become full citizens of our society without being **assimilated** by it', (in bold as cited by Molloy 1998: 7). However, the positive approach towards Travellers in 1995 evolved to the extent that equality law was passed protecting against discrimination in employment on the ***grounds*** of belonging to the Traveller Community in 1998.[9] Later, another equality law was passed in 2000 protecting against discrimination in access to services and goods, and again this was on the same grounds.[10] Nevertheless, it is important to note that Irish Travellers were ***not*** offered protection on ***racial*** grounds. Despite the optimism created circa 1998 for Travellers, as the findings and information emerging from the literature review affirmed; to this day, accommodation programmes for Travellers remain inadequate (see Irish Traveller Movement 2006).[11]

Traditionally in the Republic of Ireland, Travellers had often been subjected to control under civil law regarding illegal encampment (see ITM, 1991; McDonagh M. 2000).[12] The effects of civil law were that

---

[9] The Employment Equality Act 1998 outlaws discrimination in employment, vocational training and advertising.

[10] The Equal Status Act 2000 outlaws discrimination in access to services and goods. Other grounds covered by equality laws are: gender; marital status; family status; age; disability; race; sexual orientation and religious belief.

[11] The more positive approach towards Travellers did not lead to adequate accommodation provision. In 1995 The Report of the Task Force on the Travelling Community recommended that 3,100 new units of Traveller specific accommodation and standard housing be provided by 2000. Of these 3,100, the Task Force identified a need for 2,200 halting site bays and transient bays leaving a remainder of 900 units to be used for standard and group housing. In total, 1068 families were provided for leaving a shortfall of 2032 families without accommodation. See http://www.itmtrav.com/ (Accessed: 07/07/05).
By 2003 this shortfall had grown to 2,556 (Parliamentary question No. 384 put to the Minister for the Environment, Heritage and Local Government asking the number in total of Traveller accommodation units yet to be provided by each local authority under their respective 5 year Traveller Accommodation Programmes for 2000-2004, 18/05/04, Ref No. 14130/04; http://www.irlgov.ie/).

[12] In the past in the Republic of Ireland Travellers have been evicted from illegal encampments under the following laws: Section 31 of the 1948 Local Government (Sanitary Service) Act providing for the prohibition and exclusion of Temporary Dwellings from Local Authority areas; Section 27 of the Planning Act 1976, frequently used to seek injunctions preventing Travellers occupying caravans in many areas; Section 69 of the Road Act 1992 dealing with the erection and maintenance of Temporary dwellings and giving the power of seizure to Garda and authorised persons to remove Caravans from certain types of roads and Section 24 of the Housing Act 2003 giving power of arrest and seizure to Garda where a

Travellers were constantly threatened with eviction notices handed out by local authorities, and occasionally, private landowners, to move on to another area. Often, Travellers moved just before the date set by which they were likely to be evicted or summoned to court for failure to comply with by-laws (Joyce, 2003; DSD, 2004). Since 2002, the Housing (Miscellaneous Provisions) Act 2002, commonly referred to as the anti-trespass law, has made illegal encampment a criminal act as opposed to being merely a breach of civil law (Coulter 2002; Pavee Point 2002; Irish Traveller Movement, 2003; Lentin and McVeigh 2006). Under the 2002 Act, the control of Irish Travellers by the Irish state can be vigorous. The ramifications of the Housing (Miscellaneous Provisions) Act 2002 are that Travellers can be moved on with less than 24 hours notice. If they refuse to move on, Travellers can be arrested without the need for a warrant. Failure to comply with a Gardai's request to leave an illegal encampment can result in Travellers receiving fines up to €3000 or, being imprisoned for a month or both. Moreover, caravans and vehicles can and have been impounded (see Deegan 2006; Humphreys 2003; Traveller Visibility Group 2003) and charges are levied for each day that a vehicle or caravan is kept in storage.[13]

In 2001 during a Parliamentary debate in the Dáil and Seanad, in reply to a question (Dáil Ĕireann Volume 545, 248, 2001) [14] concerning whether or not he had met '...a delegation from South Dublin County Council to discuss the illegal traveller encampment at the Dodder[15] [in Dublin in 2001]...' the Minister for the Environment and Local Government asserted that he had and that:

> The council representatives [had] outlined the difficulties encountered by the authority as a result of the large scale unauthorised encampments by Travellers……

Caravan is parked on public or private property without consent. In effect, all areas not covered by the Roads Act 1992.

[13] See, the Housing (Miscellaneous Provisions) Act 2002.

[14] Dáil Ĕireann Volume 545, (248: 29621/01), Written Answers, Traveller Accommodation, 27/11/01, http://www.oireachtas-debates.gov.ie/D/0545/D.0545.200111270185.html (Accessed: 29/09/06).

[15] The fact that the illegal encampment at the Dodder was named in the Dáil and Seanad indicates that this incident could have been the catalyst for the introduction of harsh new anti-trespass law that was passed soon after the debate had been held.

The Minister then concluded that, '[t]he necessity for any changes in the legislative provisions regarding unauthorised encampments….will be considered' (Dáil Ĕireann Volume 545, 248, 2001).

Later, in a press release by the Irish Government on the 25th May 2002 it was claimed that:

> ….the introduction of the law [the 2002 Act] was necessary as; 'the existing powers in the housing acts to remove unauthorised encampments from public places were inadequate to deal with the large encampments that we [the Irish Government] have seen in the past year or two' (quoted in ITM 2003: 1).[16]

However, the ITM (2003: 2) criticised the Government's intention to target illegal encampments of Travellers claiming that:

> Travellers are only camped in greater numbers on public land because traditional camping areas have been closed off with boulders, low studded walls and muck heaps.

Even so, it appeared to be the case that blocking off traditional camping sites would not end the 'problem' of nomadic Travellers as created by the Irish State's failure to provide adequate accommodation provision for them (including transient sites) in the first place. In fact, as posited by Gordon (1999: 279) by creating the 'Gypsy problem', the Irish State appeared to blame Irish Travellers for their predicament thereby legitimating more discrimination and abuse, redefined as control and punishment when the 2002 anti-Trespass Act became law. What is more, in creating Irish Travellers as a problem and attempting to solve the 'problem' by control and punishment of them, the Irish Government demonstrated law's capacity for violence (Sarat and Kearns *et. al*, 1995).[17]

---

[16] The ITM attributed the quote to the following document titled, 'Government meets Traveller Organisations, Press Release from Department of the Environment and Local Government, May 2002' (ITM, 2003: 1). The actual press release could not be located.

[17] In discussion on the effects of the 2002 Act, it should be noted that the Irish Traveller Movement (2004: 11) advised that, 'In a 1997 resolution, the Sub-Commission on Prevention of Discrimination and Protection of Minorities of the U.N. Commission on Human Rights reaffirmed the right "not to be evicted arbitrarily ... from one's home, land or community." It noted that "coerced and involuntary removal" of persons from their homes and lands could result in "greater homelessness and inadequate housing and living conditions." Article 8 of the European Convention for the Protection of Human Rights and Fundamental

In Northern Ireland, the situation for Irish Travellers is slightly different. Ordinarily, citizens in Northern Ireland are protected from discrimination under equality law.[18] However, Irish Travellers are named within the Relations (Northern Ireland) Order 1997 and thus protected from wider *racial discrimination,* for example, during the eviction process from illegal camps as outlined in due course below.

The Background to Accommodation Provision for Travellers in Northern Ireland

Similar to the situation in the Republic, reflecting the beliefs of Liègeois (1994) on the social history and politics of those who are reacting to them, until summer 2006 the provision of transient sites (and other forms of accommodation) for Travellers had not been a priority for local authorities in Northern Ireland (Ellis, 2001). In fact, in 1948, akin to the situation in the Republic circa 1960, the purpose of a Government Committee had been to consider the '...problem of Gypsies and like itinerants' (Molloy 1998: 25; Noonan 1994) thereby indicating that in Northern Ireland Travellers were also perceived as a problem to be solved. A consequence of the Government Committee was that under the remit of the proposed Gypsies Bill (Northern Ireland) 1950 it was advocated that nomadic people be fined and incarcerated for camping without the consent of a landowner. Yet, the Gypsies Bill (NI) 1950 never became law due to fears expressed by local councils that it would be unworkable because of '...a slight defence against prosecution for encamping on public or common land' (Molloy 1998: 25).

Freedoms (European Convention), ensures respect for private and family life, and Article 1 of Protocol 1 to the convention protects the right to enjoyment of possessions. Often times during evictions both of these Articles are not adhered to. Examples of this include the invasion and lack of respect for a person's home (caravan/mobile) by Gardai and the impounding of the same said home'. It should be noted that the European Convention on Human Rights Act 2003 (ECHRA 2003) took effect on 12th December 2003 in the Republic of Ireland.

[18] In Northern Ireland equality law is as follows: the Equal Pay Act (NI) 1970 (as amended); Sex Discrimination (NI) Order 1976 (as amended) Fair Employment and Treatment (NI) Order 1998 (as amended); Disability Discrimination Act 1995; Equality (Disability, etc.) (NI) Order 2000 and the Employment Equality (Sexual Orientation) Regulations (NI) 2003. In addition, the Criminal Justice (No. 2) (NI) Order 2004 offers increased protection for offences aggravated by hostility; inciting hatred or arousing fear on grounds of sexual orientation or disability.

Due to the small numbers of Travellers residing in Northern Ireland in 1954, in a report entitled 'Itinerant Gypsies', the Ministry of Home Affairs (NI) concluded that, '[t]he problem of itinerancy did not warrant major legislative action or any large expenditure of public money' (in Molloy 1998: 25). During the 1950's civil law was used to evict Travellers from illegal camps and it is still used today.[19] Nevertheless, even though the numbers of Travellers residing in Northern Ireland remain low, the responses to accommodation provision for them have been and remain ineffective and as considered in due course, the Troubles could be one reason for the lack of response by local authorities towards provision of Traveller accommodation (Ellis, 2001).

In the 1960's the Caravan Act (Northern Ireland) 1963 (influenced by the earlier Caravan Sites and Control of Development Act 1960 in England and Wales) aimed to improve standards on caravan sites (Molloy 1998: 22). However, no provision was made for 'Gypsies' under the 1963 Act therefore it did not improve the situation for Irish Travellers and the situation faced by them was largely ignored until the findings of the Northern Ireland Working Party on Site Provision for Travelling People led to the sanctioning of the Local Government (Miscellaneous Provisions) Northern Ireland Order 1985. The 1985 Order was fashioned on the Caravan Sites Act (1968) in England and Wales that had heralded the misguided and later abandoned concept referred to as designation (Hawes & Perez 1995).[20] One criticism made of the Local Government (Miscellaneous Provisions) Northern Ireland Order 1985 was that it failed to take notice of the evident failures of the English and Welsh legislation that it (almost) mirrored. In fact, a problem unique to the 1985 Order was that unlike its heavily criticised counterpart (the Caravan Sites Act 1968 in England and Wales) councils were given merely the *right* as opposed to a *duty* to provide sites. The Local Government (Miscellaneous Provisions)

---

[19] The civil laws that can be used in Northern Ireland to remove unauthorised encampments are: the Public Health (Ireland) Act 1878, the Housing Act 1963, the Pollution Control and Local Government (Northern Ireland) Order 1978 and the Roads (Northern Ireland) Order 1980.

[20] Hawes and Perez, (1995) detail that in England and Wales the Caravan Sites Act 1968 specifically obliged local authorities to create spaces for up to fifteen caravans per district for Gypsy Travellers. The provision of a maximum of 15 spaces (one trailer per plot) for Traveller families was termed as designation. After this requirement had been met, local authorities were no longer obliged to consider the site provision requirements of Gypsies or Travellers remaining within their areas and this led to constant evictions for Travellers without recourse under the Act. For further information on designation see Gmelch (1982); DOE (1991); Diamondopolous (1995); Clarke (1998); DETR (1998) and Drummond (2000).

Northern Ireland Order 1985 contained no proviso for the Secretary of State to intervene and direct councils to take affirmative action for Traveller accommodation (Molloy 1998; Noonan, 1994). Hence, Travellers illegally encamped in a district that had met its quota of 15 sites (one per each family of Travellers) under the policy of designation could then be moved on and continue to face fines. The fact that the Local Government (Miscellaneous Provisions) Northern Ireland Order 1985 copied the misguided policy of designation in England and Wales, combined with its unique failings relating to intervention by the Secretary of State, entailed the consequences that, akin to the failure to provide for Travellers under the Caravan Sites Act 1968 in England and Wales (see Clements and Campbell 1997), the 1985 Order did not resolve the accommodation crisis for Travellers in Northern Ireland. In particular, one of the failings of the concept of designation was the racism inherent to it as Travellers were perceived to be:

> ....a problem/threat, to be controlled by a strategy of minimum provision, which recognises neither their nomadic lifestyle nor the natural increase in population....(Molloy 1998: 28).

A gap in official responses towards Travellers remained evident until the racism intrinsic to designation proved to be its undoing. Circa 1997 the Department of Environment realised that designation would clash with new legislation that was soon to recognise Irish Travellers as an ethnic minority.

In 1997, for the first time in any part of the United Kingdom, Irish Travellers became a notified racial group under the Race Relations (Northern Ireland) Order 1997.[21] The Race Relations (Northern Ireland) Order 1997, made it, '...unlawful to discriminate against anyone on the grounds of race, colour, nationality, or national or ethnic origin' (Mann – Kler 1997: 3). Designation would also have conflicted with Section 75[22]

---

[21] The definition of Irish Traveller is based upon terms in the Race Relations (Northern Ireland) Order 1997 explicitly referring to Irish Travellers as: the community of people commonly so called who are identified (both by themselves and by others) as people with a shared history, culture and traditions including, historically, a nomadic way of life on the island of Ireland. Thus, designation was in opposition to protection from race discrimination for Travellers.

[22] Section 75 of the NI Act 1998 came into force on 1st January 2000. It requires public authorities to have due regard to the need to promote equality of opportunity between persons of different religious belief, political opinion, racial group, age, marital status, or sexual orientation; between men and women generally; between persons with a disability and persons without; and between persons with dependants

of the Northern Ireland Act 1998 and the later Race Relations Amendment (Northern Ireland) Order 2000 both (obviously) recognising Irish Travellers as a racial group (see OFMDFM 2002). However, it should also be noted that in 1996, the Department of the Environment for Northern Ireland had established a working party to consider the accommodation needs of Travellers up until 2006. In the final report of this venture under the title: 'Accommodating Nomadism' it was revealed that, '[t]he Race Relations (NI) Order 1997 not only defines Travellers as a specific 'racial group' but identifies nomadism as a key aspect of their cultural identity.' Subsequently, it was recommended that:

> Not only should service providers ensure that they can accommodate the specific needs of Travellers who are nomadic, but policies or practices that tend to limit or hinder Traveller nomadism without good reason may actually be counted as unlawful. With this in mind, nomadism can be accommodated by: taking a strategic approach to the provision of Traveller accommodation; ensuring that specific services are adapted so that they are equally accessible to those Travellers who choose to retain a more nomadic way of life (DoE, 1999: 1).

Toleration of illegal encampments was also advised and this was influenced by the failures of legislation in England and Wales such as the 1994 Criminal Justice and Public Order Act (Clements and Campbell 1997; Thomas and Campbell 2002) to control illegal encampments amidst recognition that adequate accommodation for (especially nomadic) Travellers had not been provided.

Despite the proviso to accommodate nomadism made in the report by the DoE (1999) Lentin and McVeigh (2006: 140; see also PSNI 2002: 32) note that when the Draft Unauthorised Encampments (Northern Ireland) Order was initially considered in 2003, the consultation paper on it had been:

> …accompanied by an equality impact assessment required by Section 75 of the Northern Ireland Act.. [concluding that]…the proposal to introduce legislation to control unauthorised encampments….is likely to have an adverse impact on Irish Travellers.[23]

---

and persons without. Public authorities are also required to have regard to the desirability of promoting good relations between people of different religious belief, political opinion or racial group.

[23] Although merely a duty and not a law that can be legally enforced, under Section 75 the policies of the PSNI must be '…impact – assessed to determine if they

Despite the recommendations of the DoE (1999) outlined above and the fact that transient sites had not been developed, on 19/07/06 the Unauthorised Encampments (Northern Ireland) Order 2005[24] was passed. The provisions of the 2005 Order are almost identical to those contained within the 2002 anti-trespass Act in the Republic of Ireland and the fact that Irish Travellers are actually mentioned within the 2005 Order indicates that they remain perceived of as a problem to be solved by the state. However, relating to Section 75, due to the lack of transient site provision, if the PSNI had moved Travellers on within less than 24 hours (as stipulated within the 2005 Order) they would run the risk of being accused of racially discriminating against Travellers under the Race Relations (NI) Order 1997. Nevertheless, more recently in a debate in the House of Lords on the 20th June 2005 Lord Rooker acknowledged the accommodation difficulties for nomadic Travellers disclosing that:

> To allay concerns about the impact [of the Order] on this [Traveller] community, the [2005] Order contains a measure of protection. Accordingly, provided there are no public order issues, the police will be required to consult with the Housing Executive about the availability of an alternative suitable site when considering the eviction of Travellers from an unauthorised encampment... [and also that] I am aware that the provision of transit sites for those who pursue a nomadic lifestyle has not progressed as quickly as had been anticipated. In recognition of this, the substantive provisions of the Order will not come into effect until an adequate number of transit sites are operational (HOL 2005).[25]

---

discriminate in any way and to assess how well they promote equality on various issues including race' (PSNI 2002: 32).

[24] The Order contains nine Articles creating the power for a police officer to direct a person to leave land and remove any vehicle or other property with him on that land. It also created an offence and a power of seizure for non-compliance with a direction. In addition, the Order states that: '...any reference to the Irish Traveller community shall be construed in accordance with Article 5(2)(a) of the Race Relations (Northern Ireland) Order 1997' see Statutory Instrument 2005 No. 1961 (N.I. 11), The Unauthorised Encampments (Northern Ireland) Order 2005, http://www.opsi.gov.uk/si/si2005/20051961.htm#3 (Accessed: 10/10/06).

[25] Although the Unauthorised Encampments (Northern Ireland) Order 2005 was passed on 19/07/06, and development of transient sites has commenced, members of the public remain opposed to them, as is evident in an article appearing in *Lurgan Today,* 'Traveller Site Fury', 22/02/07, http://www.lurgantoday.co.uk/ViewArticle2.aspx?SectionID=2611&ArticleID=2064032 (Accessed: 26/02/07).

As already stated, the Unauthorised Encampments (Northern Ireland) Order 2005 mirrors the 2002 [anti-trespass] Act in the Republic of Ireland and, as indicated in Lord Rooker's speech (HoL 2005) it is indicative of a desire to enforce policies of zero tolerance against those involved in 'public order issues'. Even so, when Travellers are illegally encamped, it is highly likely that ordinarily this could be perceived of as a public order issue (see Wilson 1968: 222-3; cited in Dixon 1999: 103).[26] As such, the fact that transient sites had not been provided and yet the 2005 Order was passed is anomalous and the current (2007) situation indicates that Travellers remain blamed for their predicaments thus, attempts to punish and control them (can) appear 'legitimate'.

Policies of zero tolerance could have influenced state parties in both jurisdictions to gain greater control of the population mass including nomadic people. Equally, the ceasefire relating to the Troubles in Northern Ireland could have influenced the introduction of anti-trespass laws across the island therefore, relating to the treatment of Irish Travellers, a brief review of the social history and the politics of each jurisdiction is set out below in order to further illuminate the psychology of 'those who are reacting to them' (Liègeois, 1994: 199, cited in Gordon 1999: 291).

Policies of Zero Tolerance and Control of Travellers in the Republic of Ireland

As observed by O'Donnell and O'Sullivan (2003), even though the crime rate in Ireland had sharply decreased from 1995, statistics hid the fact that serious crime such as rape and murder had increased. Indeed, three rural murders in 1996 followed by the murder of a detective and a prominent journalist Veronica Guerin in the same year, helped to whip up a moral panic (Cohen, 1972) about crime. According to O'Donnell and O'Sullivan (2003: 48) the murders were:

> ...defining moments in the debate about law and order in Ireland. They were the catalyst for a hardening in political attitudes. The Archbishop of Dublin told mourners at Guerin's funeral that it was 'time to reflect on the drift in the direction of our society and to ask how it may be halted'

[26] It has been noted though that the term 'public order' is, '...nowhere defined and can never be unambiguously defined because what it constitutes is a matter of opinion and of convention, not a state of nature' (Wilson 1968: 222-3; cited in Dixon 1999: 103).

> (*Sunday Times, 30 June 1996*) [sic]. Crime control became a national priority and for a time it was almost as if a state of national emergency had been declared.

Subsequently, '[a] major prison-building programme was initiated, a policy of 'zero tolerance' policing was introduced, and some of the symptoms of a 'culture of control' began to emerge' (O'Donnell and O'Sullivan 2003: 41; see also Garland 1990).

The concept of zero tolerance was adopted by many nations following its success related to crime control in the city of New York from 1993 onwards (GoA 2000: 2). As stated by Fianna Fặil (1997:2 cited in O'Donnell & O'Sullivan 2003:49) the concept of zero tolerance implied that:

> …no crime, no matter how small, is insignificant; tolerance of small crimes creates a climate where big crimes flourish; and there is a certain connection between lesser and more serious crimes.

What is more, O'Donnell and O'Sullivan (2003:49) claimed that the concept of zero tolerance was:

> …appealing at a time when people [were]….sick and tired of crime, [felt]…overwhelmed by its apparently relentless upward trend and [demanded]….firm action taken to eradicate drunken loutish behaviour, intimidating gatherings of young people in public areas, the public use of obscene language, aggressive begging,[27] litter, graffiti, drug dealing and prostitution.

Typically, the targets of 'zero tolerance' were the disenfranchised '…living at the margins of society' (O'Donnell & O'Sullivan 2003:50) and the demographic profile of Irish Travellers (see for example, reports by Pavee Point 2002 and the Irish Traveller Movement 2006) indicates just how disenfranchised they are. Nonetheless, it is regrettable that policies of zero tolerance largely ignore white-collar crime (see NIO 2000: 11.23; Bauman 2000: 41; Garland *et. al* 2000; Nelken 2002). Indeed, white-collar crime is largely ignored because, '…the focus of zero tolerance is on quality of life or public order offences [regardless of the ambiguity and contentions as to definitions of them, thus] policing

---

[27] For a discussion on the effects of the Irish Children Act 2001 on Irish Travellers, begging, and the Traveller economy, see, Helleiner, J. (2003) The Politics of Traveller 'Child Begging' in Ireland, *Critique of Anthropology*, Vol 23 (1) 17-33 SAGE publications.

resources are….dedicated predominantly to those types of offences' (GoA 2000: 7).

So far, the influence of policies of zero tolerance on control of Travellers in the Republic has been considered and whilst policies of zero tolerance are implicated in anti-trespass law in Northern Ireland, undoubtedly, the Troubles have impacted on Travellers as considered below.

The Troubles and Control of Travellers in Northern Ireland

The colonisation of Ireland by the British (see for example Geraghty 1998: 247 – 354; Sexton and Kinealy 2002) has had dramatic repercussions especially for the political situation of Northern Ireland. In 1921, for many Irish people, Partition '…created a continued, unwanted British presence on Irish soil' (Geraghty 1998: 247; see also Brewer *et. al* 1996). The Troubles stem from the British colonisation of Ireland and this has undoubtedly shaped the psychology of the two major ethnic groups (see O'Rawe 2005)[28] towards each other and to outsiders that are either indigenous to, or are immigrants to Northern Ireland.

Ellis (2001) believed that from (approximately) 1968 to 1998 the Troubles (Geraghty, 1998 Sexton and Kinealy, 2002) caused both a poor economic climate and social disruption to Northern Ireland. In comparison to the rest of the UK, Partition led to a different political dynamic developing in Northern Ireland leading to distinct patterns of inter-community relations, in fact in 2001 it was observed that:

> ….more than 60% of the population [are] living in areas that have more than 80% of one religion. This residential polarization is accompanied by segregation in other areas of life such as education, religious practice, business, voluntary organisation and even in the newspapers read (Ellis 2001: 340).

Consequently, planning considerations have related to the key dispute between the Protestant/Unionist and Catholic/Nationalist communities (Ellis 2001: 340). Thus, disputes between the two major ethnic communities (O'Rawe 2005) have dominated the agenda of planning authorities to the detriment of ethnic minorities (including Travellers).

---

[28] O' Rawe (2005) describes Protestants and Catholics in Northern Ireland as ethnic communities.

During the mid 1990's, profound socio-economic changes were evident across the island of Ireland. Traditionally, due in great part to colonisation by the British, the Irish had emigrated and immigration to Ireland had been rare (Lentin and McVeigh 2006) yet, from the mid 1990's the island began to experience an influx of newcomers. Ellis (2001) observed that as the peace process developed from the early 1990's onwards, the nature of political debate in Northern Ireland and the Republic fundamentally changed to the extent that issues concerning cultural pluralism and racial equality appeared on the political agenda for the first time. In fact, it was only in 1997, the European Year Against Racism that '...the Irish state [and the jurisdiction of Northern Ireland] was suddenly compelled to start engaging...' with it (Lentin and McVeigh 2006: 1).[29] Consequently, cultural pluralism and racial equality was encouraged and guided by groups representing ethnic minorities.[30] Effectively, ethnic minority support groups drew attention to the fact that cultures existed beyond the Catholic/Protestant/Nationalist/Unionist divide, highlighting the discrimination faced by members of other ethnic groups (Ellis, 2001: 343).

According to Ellis (2001), in general, racism is more important an influence on people's attitudes in Northern Ireland towards other groups than sectarianism. More than twice the numbers of people in Northern Ireland are unwilling to mix with ethnic minorities than they would be to mix with people of the Catholic or Protestant faith (see Connolly and Keenan 2000; Brewer and Dowds 1996). However, in the case of Irish Travellers, the majority of whom *are* Catholic, the situation is markedly different. In fact, in 2000, 40 % of the people of Northern Ireland believed a Traveller lifestyle to be invalid and that it should not enjoy the support of Government whilst 57 % were unwilling to accept Travellers within their neighbourhood (Connolly and Keenan, 2000). Thus, it appears to be the case that even being Catholic will not protect Travellers from racism.

The attitudes of the majority in Northern Ireland towards Travellers appear no better than those identified in the Republic of Ireland as Brown

---

[29] The fact that the Irish state and the jurisdiction of Northern Ireland were suddenly compelled to start engaging with racism was reflected in equality laws across the island in circa 1997 that recognised Irish Travellers as an ethnic group in Northern Ireland and on *grounds* of belonging to the Traveller Community in the Republic.

[30] An example of groups supporting ethnic minorities are as follows: the Belfast Traveller Support Group (BTSG); the Chinese Welfare Association (CWA); the Northern Ireland Council for Ethnic Minorities (NICEM) in Northern Ireland and the Irish Traveller Movement in Dublin.

(2004: 5) found that 'Travellers and asylum seekers are the minorities viewed most negatively by the majority population'. Some 21 % of respondents to Brown's (2004) survey believed that Travellers should not have the same rights as the settled community, whilst 23 % believed that nomadism should not be preserved. Brown (2004: 1) also found that attitudes towards Travellers are, '[m]ore instinctive, more deeply ingrained and less subject to correction by liberal sensitivity'. Nevertheless, in Northern Ireland Brewer and Dowds (1996: 109, in Ellis 2001) found that Protestants appeared to have higher levels of prejudice against Travellers because they are perceived to be Catholic, whilst Catholic areas had the greater amount of contact with Travellers because they gravitated towards those areas. The last point is largely correct: Travellers do tend to gravitate towards Catholic areas, even so, in conversations with people belonging to either major religion across the island during my research, nearly all were surprised to learn that the majority of Travellers are Catholic, therefore the assertions made by Brewer and Dowds (1996) are perhaps spurious (see Joyce 1985: 39-40;[31] see also McCarney 2005, C: 10).[32] Overall, although it can be considered that across the island anti-trespass laws would prevent anyone from camping illegally, primarily (as exemplified by the naming of Travellers within the 2005 Order in Northern Ireland) it is Irish Travellers who commonly fall foul of and are regulated by them. Thus, via anti-trespass laws, despite lack of accommodation provision, it appears to be the case that state parties demonstrate a warped psychology as Travellers remain blamed for their predicaments thereby 'legitimising' further control and discipline of them.

So far, the background to the research has demarcated the accommodation crisis that Travellers have faced for far too long. Although a dearth of information remains on the situation of Irish

---

[31] Irish Traveller Nan Joyce (1985: 39 – 40) states that even during the Troubles, 'We'd go from the Catholic side to the Protestant side....We kept moving because we had to, there were big fines for camping.' During this research, Travellers were seen camping illegally in areas very close to Loyalist estates, therefore the claims made by Brewer and Dowds (1996) appear to be too broad and it should also be noted that Travellers remain marginalised even within the political process: none of the political parties have a charter that explicitly supports nomadism.

[32] With reference to west Belfast, McCarney (2005: 10) noted that a Traveller reeled off, '...a litany of local bars which he has been put out of, or been refused entry to, simply for being a Traveller'. As west Belfast remains clearly divided across sectarian lines it is highly unlikely that Travellers have entered bars in Protestant areas there.

Travellers (and Gypsies) with criminal justice and its agencies, the majority of research documented below on the island of Ireland, Britain and Europe, emerged prior to commencing this project and was influential in consideration of the methods.

Perceptions of Travellers and Gypsies as Congenitally Criminal[33]

The Project on Ethnic Relations (1999: 4) considered that:

> For majority populations, the concept of "Romani criminality"[34] is too often a justification for their continuing indifference or hostility toward the plight of this group. For the overwhelming majority of law-abiding Roma,[35] the burden of the criminal stereotype is one more sign that they can never get an even break in their struggle for acceptance and integration.

The consequences of stereotyping Traveller communities as deviant or criminal (Kohn 1995) are that both Traveller and sedentary communities become alienated from each other (PER 1999: 4;[36] see also Fonseca,

---

[33] The perception that a race of people could be congenitally (biologically/genetically) criminal stems from the 'science' of eugenics. It is believed that, '…the darkest period for eugenics [occurred] when Nazi Germans embarked on their "final solution" to the Jewish question, or the Holocaust'. See:http://www.georgetown.edu/research/nrcbl/publications/scopenotes/sn28.htm (Accessed: 20/03/07).

[34] As used by PER (1994) 'Romani' includes all Traveller communities.

[35] The Project on Ethnic Relations (1999) advocated that the generic term of 'Roma' be adopted by all Traveller communities throughout Europe.

[36] The report by the Project on Ethnic Relations (1999: 6- 7) considered that, '[of] all the stereotypes of the Roma, [all Gypsies and Travellers] the social construct of "Gypsy criminality" has had the most devastating impact upon the social status of the community. In public discourse concerning the Romani community - whether political or in the media - the issue of alleged Gypsy criminality…. is often referred to as a justification for society's treatment of the Roma. Deeply rooted in the past and widely held, this stereotype invariably shapes public opinion and state action against the Roma.' In addition it is stated that, '[a]nd there is no doubt that the policies and actions of the state and of lawmakers contribute greatly to the perpetuation of the criminality stigma. Crime prevention has often been the primary focus of government policies toward the Roma. In fact, stereotyping the Roma as a criminal or asocial group, as the Nazis did, was at the heart of the Porraimos, or the Romani Holocaust, during World War II' (PER 1999: 7). What is more, PER (1999: 7) comment that, '[t]he mainstream media have played a crucial, negative role in

1995).[37] The alienation between the two communities is evident in the majority of articles and academic texts concerning Travellers and Gypsies in Britain and Europe.[38]

In England and Wales it is only very recently that the Commission for Racial Equality acknowledged and began to act upon the situation of Travellers vis-à-vis racial discrimination per se. Circa 2003, the Commission began to accept that in general, '[t]he situation for Gypsies and Travellers is incomparably worse than it is for any other ethnic group…' (CRE 2003: 1). The Commission for Racial Equality also observed that:

> Whilst there is little overall research available on the experiences of Gypsies and Travellers in the criminal justice system…..Issues raised in wider studies and the work of NGOs includes the high number of racist incidents; low levels of trust in police handling of Gypsies and Travellers' cases; concerns about sentencing patterns and disproportionality in stop and search…(CRE website).

In fact, Coxhead (2004) revealed that the Police Complaints Authority Annual Report (2001-2002: 65; cited in Coxhead 2004: 4) had raised concerns that:

> Our members have noted a growing number of examples during the year of allegations of discriminatory treatment of Travellers. We are dismayed that police officers, on and off duty, continue to use offensive language and stereotypes [relating to Travellers]. Such behaviour is incompatible with ethical and respectful policing and evidence of it should invariably result in formal misconduct action.

---

perpetuating the stereotypes of the Roma, arousing mistrust, fear, and sometimes hatred. The media have eagerly reported "Gypsy crimes" and used supposedly objective police statistics about the "Gypsy crime rate" in their coverage of violent incidents involving the Roma. In so doing, the media helped to justify the violence and convert victims into perpetrators. They generalized the guilt of "criminals" to the entire Romani community'.

[37] After four years of field work that included living with Gypsy families in many European countries Fonseca (1995) concluded that the Gypsies, alongside Jewish people, are ancient scapegoats

[38] Texts by the following authors typify the alienation between Gypsies and Travellers and settled people: Scraton (1977); Grőnfors (1979); Williams (1994); Hawes & Perez (1995); Mayall (1995); Okley (1997); Gordon (2000); Save The Children (2000); Dobbs (2004); Daskalaki (2000); Tatchell (2001); ERRC (2003); Williams (2003); Giles (2004); James (2005); BBC News: A (2005): A; Guest (2005); ERRC (2005); Duvall (2005); Slater (2005); Dillon (2006); Rostas (2005); Farkas (2006).

The fact that some officers use offensive language about Travellers and stereotype them reflects the 'canteen culture' (sexist/racist behaviour and language) of policing identified by Reiner (1992), referred to by Chan (1997) and O'Hanlon (2002: 49)[39] and Ford *et. al* (2006) and as described by Guest (2005).[40] Not surprisingly, Coxhead (2004: 5) also found that:

> Gypsies and Travellers have been cited as being at the 'bottom of the pecking order of diversity' by trainers and it is this issue which may undermine *all* diversity training, by allowing 'caveats' to exist.

During the training of police officers on cultural awareness of Gypsies and Travellers, Coxhead (2004: 16; see also Oakley and Taylor 2006) identified a range of negative attitudes exemplified by officers of various ranks.[41] In order to counteract issues that seriously undermine diversity training for officers on Travellers, Coxhead (2004: 2) proposed that the lessons of The Holocaust (known as the Poraimos by Gypsies)[42] could be used by anti-racism trainers, '...as a focus for strategies to provoke both intellectual and emotional responses about genocide'.

---

[39] In research on the attitudes of police officers towards Travellers O'Hanlon (2002: 49) identified that, 'Most interviewees stated that they had experienced colleagues using derogatory terminology, though not in face-to-face contact with the Traveller people....'

[40] Guest (2005) confirmed that, as observed by a Romany police officer who was off duty and attending the Gypsy horse fair with family members at Appleby in England, fellow officers could be openly hostile towards Travellers.

[41] Examples given by Coxhead (2004: 16) of police attitudes towards Travellers are as follows: '...some officers state that all Gypsies are thieves and will not alter their views even when challenged by anti-racist trainers; reference by police to Gypsies and Travellers as thieves and scum and thieving Gypsy bastard is a common theme even among probationer constables during initial training; officer students were found to be very comfortable to be very discriminatory towards Gypsies and evidence arose of a student refusing to listen to any information on Gypsies and Travellers; police officers stated that the negative view of Gypsies is supported by fact; police officers stated they had stop checked on the sole basis that a suspect was a Gypsy and evidence arose of police officers driving around Gypsy sites doing wheel spins, flashing lights and sounding two tone horns'.

[42] In Romani the term Poraimos (also Porrajmos) translates literally as 'devouring' and was coined by Roma people to describe attempts by the Nazi regime to exterminate most of the Roma peoples of Europe during The Holocaust, see, .http://en.wikipedia.org/wiki/Porajmos, (Accessed 10/01/07). See also the web site of Professor Ian Hancock, member of the United States Holocaust Commission for the Roma and Sinti (Gypsies), http://www.chgs.umn.edu/Histories__Narratives__Documen/Roma___Sinti__Gypsies_/roma___sinti__gypsies_.html, (Accessed: 10/01/07).

Coxhead (2004: 2) believes that The Holocaust can be used '..as both a warning from history and a case study of how the 'normal moral framework' was abandoned' (see Kohn 1995; Gypsy Council 2007: 4). Indeed, as observed by Bauman (1989: xii) if society continues to view The Holocaust as an event that '…happened 'out there' – in another time, another country' then the danger remains that:

> The implication that the perpetrators of The Holocaust were a wound or a malady of our civilisation – rather than its horrifying, yet legitimate product – results not only in the moral comfort of self-exculpation, but also in the dire threat of moral and political disarmament (Bauman 1989: xii).

Although arguable, Coxhead's (2004) report indicates that 'the perpetrators of The Holocaust' were *not* 'a wound or a malady of our civilisation', (like some police today), they were in fact its 'horrifying yet (questionably) legitimate product'. Although no reports exist to suggest that police have committed murder against them, as indicated by Coxhead (2004) many police officers and their organisations remain morally and politically disengaged vis-à-vis Travellers. In fact, it appears to be the case that police can do as they please as no disciplinary procedures appear to apply in the case of the poor treatment of Travellers. Yet, even if disciplinary procedures do exist, the fact that, '…senior officers can use racist language when discussing Gypsies and Travellers' (Coxhead 2004: 16) indicates that rules and regulations are not always adhered to. As observed by Bauman (1989: xii), if the lessons of The Holocaust remain ignored, which appears to be the case relating to police and Travellers then the dangers are that the message that it contains:

> ….about the way we live today – [for example] the quality of the institutions on which we rely on for our safety…[and] the validity of the criteria with which we measure the propriety of our own conduct and of the patterns of interaction we accept and consider normal – is silenced, not listened to, and remains undelivered.

In particular then, The Holocaust contains lessons for police and society in measuring 'the propriety of our own conduct and of the patterns of interaction we accept and consider normal' (Bauman 1989: xii) concerning the treatment of Travellers (and other minorities). Despite lessons that should have been learned from The Holocaust though, O'Hanlon (2002: 5) observed that in general:

> …Gypsies and Travellers arguably [sic] continue to be excluded by their lack of "formal belonging", or their "institutional-political acceptance" by

> virtue of their invisibility in official statistics such as census records and criminal justice data.

Two decades previously, in work on Finnish Gypsies and the criminal justice system, citing Radzinowicz and King (1979) Grőnfors (1979: 15) had raised similar concerns suggesting that:

> It would be of great service to criminologists, to administrators and to public opinion in general, if national or ethnic origins, both of offenders and of victims, could be shown, along with sex and age. If differences in criminal convictions were thus brought to light, we should have a more solid foundation to allay groundless fears or as a basis for understanding, and, where possible, action.

Likewise, regarding Travellers and Gypsies, the Directorate General for Employment and Social affairs of the European Commission also identified that:

> The current gap in race-specific statistical data deprives policy makers of a clear and justifiable basis for action. Without such statistical data, little concrete information exists which might motivate change (DGE&SAEC 2004: 39).

In addition, O'Hanlon (2002: 57) observed that the, '…invisibility of the [Gypsy/Traveller] community alone [within criminal justice system statistics] suggests a potential for institutional racism…' which Macpherson (1999) defined as being, '[t]he collective failure of an organisation to provide an appropriate and professional service to people because of their colour, culture, or ethnic origin'. In fact Macpherson (1999: 6.34:39) believed that institutional racism:

> …can be seen or detected in processes, attitudes and behaviour which amount to discrimination through unwitting prejudice, ignorance, thoughtlessness and racist stereotyping which disadvantage minority ethnic people.

Subsequently, Macpherson (1999: 46.27: 39) asserted that:

> It is incumbent upon every institution to examine their policies and the outcome of their policies and practices to guard against disadvantaging any section of our communities.

In the section that follows, the lived experiences and perceptions of Travellers relating to criminal justice across Britain indicate just how disadvantaged they remain.

Travellers and the Criminal Justice System in Britain

Williams (1994) was the first author to assess the situation of Travellers within the criminal justice system in England. As a serving probation officer, Williams (1994) had previously had contact with the Traveller community in Kent and used her status as an insider to undertake research.[43] Williams (1994) found that Travellers were more likely to be prosecuted earlier in their criminal career than non-Travellers and this finding was supported by the number of cautions Travellers had received in comparison to non-Travellers (and as observed by Mc Conville *et. al* (1991), it should be noted that multi-cautions can eventually lead to convictions).

In consideration of the construction of a 'suspect population' Mc Conville *et. al* (1991: 15) suggested that this occurs through police rules rather than legal rules and that these arise from the discretion open to police. However, apart from the use of cautions, one of the other discretions available to police described by Mc Conville *et. al* (1991) includes stop and search. Stop and search was an issue that Williams (1994) focussed on. Travellers reported stop and search practices occurring so frequently that they had lost count of incidences and one told of being strip searched at the side of the road (Williams 1994: 25). Out of 21 Travellers, 15 of them had experienced a stop and search more than five times during the previous year.[44]

According to Mc Conville *et. al* (1991: 17), '...police powers bear most heavily on young...[males]....especially black males'. Similar to many young black males, the majority of young (and older) Irish Travellers lack employment opportunities (Frawley and Mongan 2002; ECNI 2006; Equality Authority 2006) and the fact that the majority lead their lives in public rather than private spaces could influence police

---

[43] Due to the absence of data on Travellers within the British criminal justice system, Williams (1994) decided to create it with the help of her colleagues. From April 1st until September 13th 1994, Kent Probation Service included an additional category of Traveller within the monitoring of pre-sentence reports. Thus, comparisons of proposals made in pre-sentencing reports and decisions taken in sentencing between black/white and Traveller populations were enabled. Probation officers also completed an additional 21 questionnaires with Travellers and interviewed six Travellers already known to them on their caseload.

[44] Williams does not present stop and search incidents related to non-Travellers in her survey. Even so, if the reader were to ask 20 white non-Traveller friends the same question regards stop and search, it is doubtful that the results would be even remotely similar.

attitudes towards them (Mc Conville *et. al* 1991). The consequences could be that agencies of control are more likely to stereotype and use judgements such as regards to 'demeanour' to identify and criminalise Travellers perceiving them to be a 'suspect population' (Mc Conville *et. al* 1991: 17).

In England as exemplified by Williams (2002) and O'Hanlon (2003) and in Scotland as demonstrated by Taggart (2003), some police officers have expressed concerns as to the treatment of Travellers within the criminal justice systems. In 2002, Nick Williams of the Diversity Directorate Unit Scotland Yard presented an unpublished paper to the Traveller Law Reform Conference in London. Praising the contribution towards anti-racism made by Traveller support groups the opinion of Williams (2002: 1) was that:

> ....without exception, the work they have carried out, in some cases over many years, has, I think, been key in finally forcing central government, local authorities and even police services to begin to wake up and recognise the appalling way in which Gypsies and Travellers have been treated in the past, and continue in many cases to be treated today.

Even though information regarding the experiences of Travellers and criminal justice remains rare, what is well known is that certain ethnic groups are over-represented within criminal justice systems across Britain and Europe.[45] However, there remains a lack of information on the experiences of indigenous people and criminal justice. Subsequently, progressive stances taken by officers such as Williams (2002) towards Travellers demonstrate a willingness to engage in further dialogue which in itself may encourage criminological research on indigenous minorities and policing.

Williams (2002) considered that lack of trust of police by Travellers is especially important relating to reporting and investigating hate crimes (Tatchell 2001; Burney, 2002; Jacoby 2002). With regard to racially aggravated cases, Burney (2002) advocated that '...however slight the charge and the sentence, prosecution symbolises the importance of confronting manifestations of racism' (Burney, 2002: 108, in Iganski, 2002). Tatchell (2001: 58) also proposed that, '[t]o preserve harmonious community relations, it is therefore right that hate crimes should attract more severe penalties'. Alternatively, even though Jacoby (2002: 121)

---

[45] For example, see Mc Conville, Sanders, & Leng (1991); Fitzgerald and Sibbitt (1997); Henderson and Cullen (1997); Chan (1997); Reiner (2000); NACRO (2000); Sanders and Young (2002); Phillips and Bowling (2003); Leung *et. al (*2004); Comack (2004); Menjivar and Bejarano (2004); Hayes, (2005).

observed that critics of hate crime legislation denounce it as 'Orwellian'[46] punishment for unorthodox thoughts, the killing of an Irish Traveller boy in England demonstrated the extremes of hate crime and the priority that police (and society in general) should give to confronting and challenging it. As revealed by Coxhead (2004: 4):

> The tragic murder of Johnny Delaney [a young Irish Traveller] at Ellesmere Port in Cheshire in 2003 shows stark testimony of the very real existence of hate crime towards Gypsies and Travellers. [The murder of] Johnny Delaney should remind us all as criminal justice practitioners that not enough is being done, and that we must act now to do more.

Nevertheless, Williams (2002: 11) voiced concerns that:

> Unless police forces can win the trust and confidence of Gypsies and Travellers......we will never get people to report hate crime and unless it is reported we can never effectively respond to it, it's a vicious circle.

Alternatively, O'Hanlon (2002) confirmed that police can be equally wary of Travellers.

As an insider, O'Hanlon (2002) was in a very advantageous position to review the perceptions and experiences of fellow officers on Gypsies and Travellers.[47] O'Hanlon (2002:40) identified that the overwhelming response of officers was that '...there is a lack of dialogue with Traveller communities, which was perceived to be based on mutual prejudice and distrust'. Two officers suggested that in contrast to searches of non-Travellers, searches of Travellers were likely to be more thorough, based on pervasive negative stereotypes and the low likelihood of sanctions through the complaints procedure, in fact, police officers perceived Gypsies and Travellers as not being politically inclined towards litigation. Two interviewees also suggested that any reluctance to stop Travellers was more likely to be due to the hostility or violence of Travellers anticipated by police. Nevertheless, the fact that some Travellers could be hostile towards police indicates what Reiner (2000: 126) described as 'interactional discrimination', a phenomenon more regularly discussed in

---

[46]Jacoby (2002) is concerned that taken to extremes, individuals could be punished by 'thought police', for example, family members or friends could report relatives for expressing racist views on others in the privacy of their own home.

[47] O'Hanlon (2002) interviewed 7 police officers with work experience ranging from 7 - 28 years in length. Using semi-structured interviews O'Hanlon (2002) interviewed custody officers; uniformed response officers; Gypsy Traveller Liaison Officers and staff who had been involved in policy-making related to Gypsy Travellers.

relation to 'the hostility of young Black men towards the police, borne out of the long standing prejudice directed towards them' (*ibid*: 126). Interactional discrimination could manifest itself in disproportionate levels of arrest and convictions of Travellers and this was a point raised by one of O'Hanlon's (2002: 47) interviewees. O'Hanlon (2002: 49) also observed that:

> Two interviewees suggested that complaints and calls for assistance [from Travellers] had not received the level of response afforded to other members of society.....

In addition to the reports outlined so far, two other authors indicated that Travellers are considered to be a suspect community within the criminal justice system. Power (2003) insinuated that Travellers were more likely to receive custodial sentences than community/probation orders or bail[48] and Bower (2005) found that Travellers have little confidence in police, advising Travellers to report crimes against them (and others) to third parties.[49] Similarly, in the following section it is evident that lack of trust between Travellers and police across the island of Ireland remains an issue.

---

[48] According to a probation officer interviewed by Power (2003) Travellers were more likely to receive custodial sentences because officials perceived that, due to the nomadic elements of their lifestyle, Travellers would be unable or unlikely to be compliant with other orders. Even if a Traveller defendant was living in a house for years, a probation officer suggested that to reveal the individuals' nomadic past would be enough reason for courts to refuse bail and remand them in custody. Along similar lines, a probation officer interviewed by Power (2003: 20) was unsure of the value of elaborating on a Traveller offenders' life style in court because the probation officer believed that agencies of control are ignorant of or prejudiced against Traveller culture.

[49] The following quote from Bower's (2005) CD rom indicates Travellers' experiences with police in Cambridgeshire in England: '[n]o way on this earth would I report anything [to police]...I would rather get our own family to sort it out because police would never do anything..... I guarantee you now if I went up to the police station and said excuse me, please do something about them [racists] they wouldn't even bother picking up a bloody phone. And it's not because they don't believe in what I am saying, it's because [of] who I am (Traveller woman)'.

Irish Travellers and Criminal Justice across the Island of Ireland

Only one report on Travellers and criminal justice in Northern Ireland emerged from the literature review (Boyle 2005). Mainly, Boyle (2005) found the attitudes of Travellers to be negative towards the Police Service for Northern Ireland and the criminal justice system in general. The only positive found by Boyle (2005) was in relation to the probation service where one young Traveller had reported a good experience of working with an officer. However, due to discrimination by potential employers, the probation officer was unable to find a work placement for the Traveller.

Travellers felt powerless to challenge the authority of the criminal justice system and it was found that they fear making a complaint against the police and/or the criminal justice system due to perceptions that if they do complain, authorities will harass them even more (en masse) in future. Although Travellers were not aware of the existence of any kind of procedures through which complaints against criminal justice agencies could be made, many were opposed to Traveller support groups taking part in police district partnerships. Boyle (2005) also identified that when in prison some Travellers have been bullied by officers and inmates alike; that there is a lack of training on Travellers within all areas of the criminal justice system and that the training that did exist appeared to be tokenistic. In fact, Boyle (2005) found that district inspectors or officers have training in Traveller awareness only on one day a year and that even though on occasions Travellers attended these sessions and gave presentations lasting an hour or two that was not considered adequate, especially as officers are then meant to feed that training back to their staff. Boyle (2005) also found that Travellers are not ethnically monitored, and that there is a pressing need to create the position of a legal adviser in order to enable liaison between them and the criminal justice system.

Some information relating to Travellers and criminal justice appeared long after the data collection process had ended. One issue of major importance to the debate on the alleged criminality of Travellers arose in the findings of the Morris Tribunal 2006. Somewhat reflecting Bauman's (1989: xii) observations on the lessons to be learned from The Holocaust and the subsequent threat of moral and political disengagement, the Morris Tribunal (2006) found that Detective Sergeant John White had:

> …deliberately planted a shotgun at a Travellers' encampment, [at Burnfoot in County Donegal] something the State [had] failed to prove beyond a reasonable doubt in a recent criminal prosecution' (Irish Times

Comment 2006: 17).

The following statement was taken from a colleague of White's presented to the Morris Tribunal (2006):

> When we got to the edge of the farm buildings, approximately 150 - 200 yards from the [Traveller] site I felt sick. I felt like a criminal. I felt everything in my head was telling me I shouldn't be here, to get out, but I felt it had gone too far and I had no control of events. The dogs started barking at us but [White] he said "keep your eyes open" and he stepped into the shadows of the buildings. When he emerged I knew he had planted the gun. We made our way back to the car. He was happy with the operation and all I could hope for was that the real gun would be found and that there would be no need to find this one, the planted one (Morris Tribunal 2006, Findings on Burnfoot).[50]

After planting a gun at the Traveller site at Burnfoot County Donegal Detective Sergeant John White had organised a search of the site in order that guns would be found and seven Travellers were arrested on suspicion. Whilst held in custody, the Morris Tribunal (2006) found that Travellers were, '....referred to casually by derogatory names....(Gartland 2006: 7). Subsequently, the Morris Tribunal 2006 (see also O'Keeffe 2006 Irish Examiner 2006: E; Irish Examiner 2006:F; Ireland Online 2006: A) indicated the inherent dangers for Travellers relating to prejudice within the criminal justice system; that they can become scapegoats for crime per se and that in some instances that they can become scapegoats for crime committed by police officers.

After the data collection for this research had finished, another important report emerged by Mulcahy and O'Mahony (2005). Mulcahy and O'Mahony (2005) interviewed 40 people, including police officers and conducted 10 focus groups with Travellers. Not surprisingly, the research by Mulcahy and O'Mahony (2005) reflected the experiences of Travellers in Britain. In fact, Mulcahy and O'Mahony (2005: 23) found that '...much of the interaction between the Traveller community and the police is in a negative context and this makes it difficult to develop any degree of trust between both groups'. Their findings indicated that Travellers are over-policed with one Traveller having experienced that police either '...lean towards [a] heavy-handed policing [of Travellers], or a kind of hands-off approach' (Mulcahy and O'Mahony 2005: 23).

[50] To see the report on the arrest and detention of seven persons at Burnfoot, County Donegal on the 23rd of May 1998 and the investigation relating to the incident, search for 'Burnfoot' on the Morris Tribunal website.

Moreover, it was found that, '…many Travellers believe that the police view any contact with them as an opportunity to trawl for evidence of illegal behaviour' (Mulcahy and O'Mahony 2005: 2). It was also suggested that, '[t]he experience of being 'moved on', in particular, [sic] is at the core of Traveller identity, as is the role of the Garda Síochána in this process' (*ibid:* 2). Travellers also complained that:

> …the police do not respond to their requests for help, whether in relation to reporting criminal matters or other issues…[and]….when they do respond, they fail to intervene appropriately (Mulcahy and O'Mahony 2005: 26).

Relating to training police in awareness of Traveller culture, as found by Coxhead (2004) in England, Travellers who had taken part in such sessions raised specific issues recalling, '…a great deal of hostility towards Travellers' among recruits (*ibid:* 26). Overall, Mulcahy and O'Mahony (2005: 23) concluded that:

> …our research suggests that relations between Travellers and the police are extremely difficult and contentious……One Traveller described the relationship between Travellers and Gardai as based on control, containment and harassment.[51]

Conclusion

The literature informing this research project exemplified that '…whenever we look at how Gypsies and Travellers are treated' we are indeed able to consider the social history and the politics of those who are reacting to them. In both jurisdictions, recent history showed that provision of transient sites for Travellers has never been a priority for either state across the island. Circa 1950's - 60's in both jurisdictions, Irish Travellers, regarded as itinerants, were perceived to be the problem, the answer being to absorb them into the general community. In general, the Irish were concerned to 'expunge Tinkerism [from Irish society] not to vindicate or assess its significance' (Court 1985: 1).

Although much later, Travellers were recognised on the grounds of

[51] Reasons for the alleged high incarceration rate of Irish Travellers in Irish jails (in contrast to those of other white Western Europeans) are given in work by the following authors: Linehan et. al (2002) Duffy et. al (2006) Wright et. al (2006) whilst Fountain's (2006) work considers the issue of drug use by Travellers.

belonging to the Traveller Community in the Republic of Ireland and as an ethnic and racial group in Northern Ireland, within both jurisdictions, state parties introduced harsh new anti-trespass laws across the island. The introduction of anti-trespass laws indicates that Travellers remain conceptualised by state parties as a problem to be solved.

With further relevance to what the social history and the politics of those who react to Travellers can indicate, it was suggested that policies of zero tolerance could have heavily influenced and shaped Irish legislation. Indeed, policies of zero tolerance could explain the power behind harsh new controls of nomadic Travellers. However, relevant to the social history of Northern Ireland, it was discussed how in 1921, for many Irish people Partition '...created a continued, unwanted British presence on Irish soil' (Geraghty 1998: 247). What is more, in comparison to the rest of the UK, Partition led to a different political dynamic developing in Northern Ireland and as such, key disputes between the Protestant/Unionists and Catholic/Nationalist communities could explain the slow response to Traveller accommodation provision there. The Troubles could also explain the psychology of those who react to Travellers in Northern Ireland as suggested by Ellis (2001) and other authors. However, a disdain for Travellers appears to remain equal in both jurisdictions. The fact that the majority of the public appear to remain opposed to the integration of Travellers could continue to have a profound influence on policy developments related to provision of transient sites and other accommodation for Travellers (as indicated by the Lurgan Today article in 2007).

Overall, the initial literature review indicated that, as considered by the Project on Ethnic Relations (1999: 4), '[f]or majority populations, the concept of "Romani criminality"' is too often a justification for their continuing indifference or hostility toward the plight of this group'. In addition, whatever they do and however they act Irish Travellers remain constructed as a 'problem' by state parties. What is more, it is evident that some of the lessons of The Holocaust have indeed been forgotten. As observed by Bauman (1989: xii), if society continues to view it as an event that '...happened 'out there' – in another time, another country' then the danger remains that we are morally self-exculpated to do as we see fit with Travellers and Gypsies (and other minorities); the consequences being that (apparently, whenever it suits) we become both morally and politically disengaged from the inhumane treatment of certain groups.[52]

---

[52] Ultimately, the findings in my thesis indicated that Irish Travellers remain perceived of as a criminal fraternity and that on occasions this situation is

References

Bauman, Z. (1989) *Modernity and the Holocaust*, Polity Press, Cambridge.

—. (2000) 'Social Uses of Law and Order', Chpt 2: PP: 24 – 45, in Garland, D; Sparks, R.(Eds.) (2000) *Criminology and Social Theory*, Oxford University Press, Oxford.

BBC News, (2005: A) 'Gypsies are 'Europe's Most Hated', 26/04/05, http://news.bbc.co.uk/1/hi/uk/4486245.stm, (Accessed: 24/07/06).

Bowers, J. (2005) DVD, Del *Gavvers Pukker-Cheerus* (Romany for: give the police a chance) Advice from Gypsies and Travellers on sorting out racists, presented by Jake Bowers, produced by The Gypsy Media Company, funded by Home Office, BCU and Central Division of Cambridgeshire Constabulary.

Boyle, M. (2005) *Travellers and the Criminal Justice System in Northern Ireland, University of Ulster*, Magee, MS.c Professional Development in the Community, (unpublished).

Brewer, J.D.; Guelke, A.; Hume, I.; Maxon – Brown, E. Wilford, R. (1996) *The Police: Public Order and the State,* Macmillan Press Basingstoke.

Brown (2004) 'Know Racism Campaign', *http://www.knowracism.ie/index.html,* (accessed 12/06/05).

Chan, L.B. (1997) *Changing Police Culture, Policing in a Multicultural Society,* Cambridge University Press.

Clarke, N. (1998) 'Living on the Edge of Your Town: A Study of Traveller Provisions and Conditions in Outer London and the Home Counties', (Eds. K. Egan, M. Wake) *BRIDGE*, Spider Pubs London.

Clements, L. and Campbell, S. (1997) 'The Criminal Justice and Public Order Act and its Implications for Travellers' P: 61-69, in T. Acton (ed) (1997) *et. al, Gypsy Politics and Traveller Identity, University* of Hertfordshire Press.

Cohen, S., (1972) *Folk Devils and Moral Panics: The Creation of the Mods and Rockers,* MacGibbon and Kee, London.

Comack, E. (2004), 'The Gendering and Racializing of Criminalized Others', p: 325- 329, in: *The Blackwell Companion to Criminology,* C. Sumner, (ed), Blackwell, London.

Commission for Racial Equality England and Wales (2003) *Gypsies and Travellers – A Strategy 2003 – 2006,*

---

exacerbated by the actions and attitudes of certain members of the criminal justice systems across the island of Ireland.

http://www.cre.gov.uk/media/nr_arch/2003/nr031027.html, (Accessed: 20/07/04).

Connolly, P. Keenan, M. (August, 2000) *Opportunities For All,* Northern Ireland Statistics and Research Agency.

Coulter, C. (2002) 'Constitutionality of Trespass Law Queried', *Irish Times,* 24/07/02, P: 4.

Coxhead, J (2004) *'Anti-Racism Training Strategies for the Police Service and Criminal Justice Sector, Gypsies and Travellers',* Paper presented to the Police Service and Criminal Justice Sector, 01/08/04, Derbyshire Police Headquarters.

Daskalaki, I. (2000), 'Greek-Gypsy identity and the relationship between Greek-Gypsies and the state', *http://www.lse.ac.uk/collections/hellenicObservatory/pdf/symposiumPapersonline/Daskalaki.pdf,* (Accessed: 26/07/06).

Deegan, G. (2006), 'Travellers Caravan's are Seized', *Irish Times,* 27/08/06, P:6.

Department of the Environment for Northern Ireland, (1999) 'New Policy for Accommodation on Travellers: Report of the PSI Working Group on Travellers', *http://www.newtsnni.gov.uk/consultation/recommendations2.htm* (Accessed: 12/05/06).

Department of The Environment, Transport and the Regions (1998) *Managing Unauthorised Camping: A Good practice Guide*, Home Office London.

Diamondopolous, (1995) 'Gypsy Traveller Information', Buckinghamshire Gypsy Traveller Project.

Dillon, E. (2006) *The Outsiders: Exposing the Secretive World of Ireland's Travellers,* Merlin Publishing, Dublin.

Directorate General for Employment & Social affairs of the European Commission, (2004), *The Situation of Roma in an Enlarged European Union; Fundamental Rights and Anti-discrimination*, Luxembourg: Office for Official Publications of the European Communities.

Dobbs, M. (2004) 'Invasion of the Travellers', *The Daily Mail*, Irish edition, 06/11/04, P: 2.

Drummond, A. (2000) *The Effects of Legislation upon Gypsies and Travellers within the United Kingdom,* Undergraduate Dissertation at Anglia Polytechnic University, Cambridge Campus.

Duffy, D., Linehan, S., Kennedy, H.G., (2006) 'Psychiatric Morbidity in the Male Sentenced Irish Prisons Population', *Irish Journal of Psychiatric Medicine*, 23(2): 54-62

Duvall, M. (2005) 'Ethnic Profiling by Police in Europe Evidence of

Ethnic Profiling in Selected European Countries', p. 14- 26, in: *Justice Initiatives, Ethnic Profiling by Police in Europe,* The Open Society Justice Initiative, http://www.soros.org/resources/articles_publications/publications/justiceinit_20050610/justiceinit_200506.pdf, (Accessed: 29/07/06).

Ellis, G. (2001) 'The Difference Context Makes: Planning and Ethnic Minorities in Northern Ireland', *European Planning Studies,* Vol. 9, No. 3.

Equality Authority (2006) *Travellers and Ethnicity*, Equality Authority Dublin.

Equality Commission for Northern Ireland, (2004) 'Response to the Department of Social Development's Draft Unauthorised Encampments (Northern Ireland) Order 2004,' *http://www.equalityni.org/uploads/word/FinCommRespUnauthorisedEncampmentsOrder131204.doc,* (Accessed: 07/05/06)

Equality Commission for Northern Ireland, (2006, A) Mainstreaming Equality of Opportunity for Travellers in Education: Towards a Strategy, ECNI, Belfast.

Farkas, L. (2006) 'The Monkey that does not See', European Roma Rights Centre, *http://www.erc.org/Popup_index.php?type=nyomtat&id=1940*, (Accessed: 10/10/06.

FitzGerald M. & and Sibbitt, R. (1997) *Ethnic Monitoring in Police Forces: A Beginning,* Home Office Research Study 173, a Research and Statistics Directorate Report, Home Office, London.

Fonseca, I. (1995) *Bury Me Standing.* Chatto, London.

Ford, R. Woolcock, N.; Tendler, S. (2006) 'Canteen Culture Leaves a Bad Taste for Many', *The Times,* 16/01/06, http://www.timesonline.co.uk/article/0,,2-1987574,00.html., (Accessed: 20/01/07).

Fountain, J. (2006) *National Advisory Committee on Drugs 'An Overview of the Nature and Extent of Illicit Drug Use Amongst the Traveller Community: an Exploratory Study,* Published by the Stationery Office, Dublin.

Frawley, O. & Mongan, M. (2002) Travellers' Accommodation Needs Assessment in Northern Ireland, Research Unit Northern Ireland Housing Executive.

Garland, D. (1990) *Punishment and Modern Society, A Study in Social Theory*, Clarendon Paperbacks.

Garland, D; Sparks, R.(Eds.) (2000) *Criminology and Social Theory*, Garland, D & Sparks, R.; Bauman, Z.; Braithwaite, J.; Cain, M.;

Lianos, M. & Douglas, M.; Hirst, P.; Melossi, D.; Rose, N. Oxford University Press, Oxford.

Gartland, F. (2006) 'Garda 'planted gun' in Traveller Encampment', *The Irish Times,* Morris Tribunal Reports, P: 7.

Geraghty, T. (1998) *The Irish War: The Military History of a Domestic Conflict*, Harper Collins London.

Giles, T. (2004) 'Gypsies Tramps and Thieves?' School of International Public Affairs, Columbia University, *http://www.columbia.edu/cu/sipa/PUBS/SLANT/SPRING/gypsies.html* (Accessed 28/09/04).

Gmelch, G. and Gmelch, S. (1974) 'The Itinerant Settlement Movement: Its Policies and Effects on Irish Travellers', *Irish Studies,* Vol. 68, pp.1-16, Spring 1974.

Gmelch, S. (1975) Tinkers and Travellers, Dublin: O'Brien Press.

Gmelch G. and Gmelch S. (1976) 'The emergence of an ethnic group: The Irish tinkers', *Anthropological Quarterly,* Vol. 49(4), pp.225-238

Gmelch, S. (1985) 'Gypsies in British Cities: Problems and Government Response', in *Urban Anthropology*, Vol. 2, pt 3 – 4, PP: 183 – 96.

Gmelch, G.(1985) The Irish Tinkers: the Urbanization of an Itinerant People (2nd ed) Prospect Heights, IL: Waveland Press.

Gmelch, S. (1989) 'From Poverty Sub-culture to Political Lobby: The Traveller Rights Movement in Ireland' in C. Curtin and T. Wilson (eds.) *Ireland from Below*, Galway University Press

Gordon, E., (2000), *The Nomadic Scapegoat: The Criminalisation and Victimisation of Gypsies*, (BL) B5d Ph.D, University of Wales, Bangor, 49-10230.

Government of Australia (2000) 'Discussion Paper, Zero Tolerance Policing, its Background and Implications for Aboriginal People'.

Grőnfors,M. (1979) *Finnish Gypsies and the Police: An Examination of a Racial Minority and its Relationship with Law Enforcement Agents,* London School of Economics and Political Science, University of London.

Guest, K. (2005), 'Romany constable blows the whistle on police prejudice towards travellers', *Independent on Sunday,* 01/05/05, P:2.

Gypsy Council (2007) The Holocaust, P: 4, *The HUB, Newsletter of the Gypsy Council,* Winter 2007, UK.

Hawes, D. Perez, B (1995) *The Gypsy and the State: The Ethnic Cleansing of British Society*, SAUS, University of Bristol.

Hayes, B. (2005) 'A Failure to Regulate: Data Protection and Ethnic Profiling in the Police Sector in Europe', p. 32-43, in: *Justice Initiatives, Ethnic Profiling by Police in Europe,* The Open Society

Justice Initiative, http://www.soros.org/resources/articles_publications/publications/juticeinit_20050610/justiceinit_200506.pdf, (Accessed: 29/07/06).

Helleiner, J. (1993) 'Traveller Settlement in Galway City: Politics, Class and Culture', in C. Curtin, Hastings & T.M. Wilson (Eds) *Irish Urban Cultures,* Queen's University, Belfast.

Henderson, M.L.; Cullen, F.T. (1997) 'The Impact of Race on Perceptions of Criminal Injustice', *Journal of Criminal Justice*, Vol. 25. No. 6, pp 447-462.

House of Lords (2005) 'Debates: Unauthorised Encampments (Northern Ireland) Order', 20/06/05, http://www.theyworkforyou.com/lords/?id=2005-06-20b.1478.6, (Accessed: 20/01/07).

Humphreys, J. (2003), 'When a Child's Bed is a Car Boot: Increasing numbers of Travellers Being Evicted', *Irish Times*, 19/07/03, P: 53.

Ireland Online, (2006: A), 'More Damning Morris Reports to be Made Public', 15/05/06, http://breakingnews.iol.ie/news/story.asp?j=182724096&p=y8z7z48xz (Accessed: 16/05/06).

Irish Examiner (2006: E) 'Bid to Frame Travellers followed by Cover-up and Lies', 17/08/06, http://www.irishexaminer.com/breaking/story.asp?j=192492312&p=y9z493xy8&n=192493072, (Accessed: 09/09/06).

—. (2006:F) 'Minister in Bid to Block Reports' Release', 22/05/06, http://www.irishexaminer.com/breaking/story.asp?j=183434332&p=y83435x38&n=183435092, (Accessed: 23/06/06).

Irish Traveller Movement (1991) Traveller Accommodation and the Law, The Report of an ITM Seminar held on 21s September 1991 at Mary Immaculate College, Limerick.

—. (2003) 'An Analysis of the use of the Housing (Miscellaneous Provisions) Act 2002 Forced Eviction' *http://www.Irish Traveller Movementtrav.com/*

—. (2003) 'Irish Traveller Movement Traveller Legal Unit, Strategic Plan 2003 – 2006'.

—. (2004) 'A Briefing document on the use of the Housing (Miscellaneous Provisions) Act, 2002', *http://www.itmtrav.com/pdf/TresspassPaper.pdf#search='Housing%20Miscellaneous%20Provisions%20%20Act%202002,* (Accessed: 10/10/06).

—. (2006: A) 'Broken Promises, An analysis of Traveller Accommodation Provision', *http://www.itmtrav.com/accommodation.html* (Accessed:

20/01/07).

Jacoby, J. (2002) 'Punish Crime, Not Thought Crime,' pp: 104-123. in: P. Iganski, (Ed.) Burney, E.; Jeness, V.; Lawrence, F.M.; Levin, J., Phillips, M.; Ray, L.; Smith, D.; Tatchell, P., *The Hate Debate: Should Hate be Punished as a Crime?,* The Institute for Jewish Policy Research, London.

James, Z. (2005) 'Eliminating communities? Exploring the implications of policing methods used to manage New Travellers', *International Journal of the Sociology of Law,* 33 (2005) 159–168.

Janicek, K. (2006), 'Gypsy Women Confront Czechs with an Ugly Legacy', *The Boston Globe*, 18/06/06, http://www.boston.com/news/world/europe/articles/2006/06/18/gypsy_women_confront_czechs_with_an_ugly_legacy/, (Accessed: 20/06/06).

Joyce, N. and Farmar, A.(1985) *My Life on the Road: A Travelling Woman's Autobiography*, A & A Farmar publishers, Dublin.

Joyce, D. (2003) Irish Traveller Movement Traveller Legal Unit Presentation to The Burren Law School; '*Outsiders and the Law, Including Those Forced Out by Disability, Ethnicity or Stigma,'* 13/05/03, Newtown Castle, Ballyvaughan, Co. Clare.

Kohn, M (1995) *The Race Gallery: The return of Racial Science,* Cape pubs. London.

Lentin, R. & McVeigh, R. (2006) *After Optimism? Ireland, Racism and Globalisation,* Metro Ėireann Publications, Dublin.

Leung, A., Wolley, F., Tremblay, R.E., Vitaro, F. (2004) 'Who Gets Caught? Statistical Discrimination in Law Enforcement', *The Journal of Socio – Economics,* 34; PP; 289 – 309.

Linehan, S., Duffy, D., O'Neill, H., O'Neill, C., Kennedy, H.G. (2002), 'Irish Travellers and Forensic Mental Health', *Irish Journal of Psychological Medicine,* 2002; 19(3): 76-79.

Mac Laughlin, J. (1998) 'The Political Geography of Anti-Traveller Racism in Ireland: the Politics of Exclusion and the Geography of Closure', *Political Geography,* 17, No. 4 pp. 417-435, Pergamon.

Mann – Kler, D. (1997) 'Out of the Shadows, An Action Research Report into Families, Racism and Exclusion in Northern Ireland' Published by Barnardos, Belfast.

Mayall, D. (1995) *English Gypsies and State Politics,* Gypsy Research Centre, University of Hertfordshire Press.

McDonagh, M. (2000) 'Nomadism', PP: 33 – 46, in E. Sheehan (Ed) *Travellers: Citizens of Ireland,* Parish of the Travelling People, Dublin.

McDonagh, N. (2001) Travelling People, PP.: 23 – 30 in, *Irish Voices Irish Lives,* Pub. Parish of the Travelling People, Dublin.

McVeigh, R. (1997) 'Theorising Sedentarism: the Roots of Anti-nomadism', p: 1- 25, in T. Acton (Ed) *et. al* (1997) *Gypsy Politics and Traveller Identity,* University of Hertfordshire Press.

Macpherson of Cluny, Sir William (1999) *The Stephen Lawrence Inquiry: Report of an Inquiry by Sir William Macpherson of Cluny*, London: HMSO

Mc Conville, M. sanders, A. Leng, R. (1991), *The Case for the Prosecution,* Routledge, New York.

Menjivar, C. & Bejarano, C.L. (2004) 'Latino Immigrants' Perceptions of Crime and Police Authorities in the United States: A Case Study from the Phoenix Metropolitan Area', *Ethnic and Racial Studies,* Vol. 27, No. 1 January 2004 PP: 120 – 148.

Molloy, S. (1998) 'Accommodating Nomadism Report', Published by Traveller Movement (NI).

Mulcahy, A.; O'Mahony, E.; (2005), 'Policing and Social Marginalisation in Ireland', *Combat Poverty Agency Working Paper* 05/02/05.

NACRO (2000) 'Race and Prisons: A snapshot survey'.

National Traveller Accommodation Consultative Committee (2004) 'Review of the Operation of the Housing (Traveller Accommodation) Act 1998', Report by the National Traveller Accommodation Consultative Committee to the Minister for Housing and Urban Renewal, Dublin.

Nelken, D. (2002) 'White-Collar Crime', Chpt 23. PP: 844 – 873, in, *The Oxford Handbook of Criminology,* (Eds. M. Maguire, R. Morgan & R. Reiner) 3RD edition, Oxford University Press

Noonan, P. (1994) Travelling People in West Belfast, Traveller Movement (NI).

Northern Ireland Office, (2000) 'Review of the Criminal Justice System in Northern Ireland'.

Oakley, R. and Taylor, C. (2006) 'Policing and Roma: A Resource Manual', Phillips, (ed), Office for Democratic Institutions and Human Rights, http://www.europeandialogue.org/reports/resource_manual.doc#Britain_training (Accessed: 23/03/07).

O'Donnell, I. & O'Sullivan, E. (2003) 'The Politics of Intolerance-Irish Style', *British Journal of Criminology,* 43, 41-62.

Office of the Deputy Prime Minister, (2005), 'Government Response to the ODPM Select Committee's Report on Gypsy and Traveller Sites', Home Office, London

Office of the First Minister and Deputy First Minister, (2002) 'A Response to the PSI Working Group Report on Travellers', Northern Ireland.

O'Hanlon, M. (2002) *An Investigation into how the Metropolitan Police Service Performs its Function in Relation to Gypsy and Traveller Communities,* MSc. Race and Ethnic Relations, Birkbeck College, University of London.

O'Keeffe, C. (2006) 'Conroy Rejects Tribunal Finding over Loss of Discipline', *Irish Examiner*, 02/09/06, http://www.irishexaminer.com/irishexaminer/pages/story.aspx-qqqg=ireland-qqqm=ireland-qqqa=ireland-qqqid=12353-qqqx=1.asp, (Accessed: 09/09/06).

Okley, J. (1997) 'Non-territorial Culture as the Rationale for the Assimilation of Gypsy Children', *Childhood: A Global Journal of Child Research*, Vol. 4, 63-80.

O'Rawe, M. (2005) 'Ethnic Profiling, Policing, and Suspect Communities: Lessons from Northern Ireland', p. 88-98, in *Ethnic Profiling by Police in Europe*, The Open Society Justice Initiative, http://www.soros.org/resources/articles_publications/publications/jutic einit_20050610/justiceinit_200506.pdf, (Accessed: 29/07/06).

Pavee Point, 2002, *Pavee Feens Hawken,* (Traveller Men Working), Traveller Economy Sectoral Partnership Project, Annual Report, Pavee Point, Dublin.

Phillips & Bowling (2003) 'Racism Ethnicity and Criminology, Developing Minority Perspectives', *British Journal of Criminology,* 43, 269-290.

Police Service for Northern Ireland (2002) *New Directions, A Human Rights and Racial Equality Focus on Policing with Ethnic Minority Communities*, Ramada Hoel, Belfast, 21st – 22nd October 2002, PSNI Northern Ireland.

Power, C. (2003), 'Irish Travellers: Ethnicity, racism and pre-sentence reports', *The Journal of Community and Criminal Justice,* NAPO, Vol 50(3): 252–266.

Project on Ethnic Relations, *'Roma and the Law: Demythologising the Gypsy Criminality Stereotype,'* Paris, France, October 7-8, 1999 Copyright 2000, http://www.per-usa.org/reports/RomaandtheLaw00.pdf

Reiner, R., (1992) *The Politics of the Police,* 2nd Edition, Sussex, St Martins Press.

Rostas, I. (2005) 'ID Checks and Police Raids: Ethnic Profiling in Central Europe', p. 26-30, in: *Justice Initiatives,* 'Ethnic Profiling by Police in Europe', The Open Society Justice Initiative, http://www.soros.org/resources/articles_publications/publications/justiceinit_20050610/justiceinit_200506.pdf, (Accessed: 29/07/06).

Sanders, A. & Young, R. (2002) 'From Suspect to Trial', *The Oxford Handbook of Criminology,* (Eds. M. Maguire, R. Morgan & R. Reiner) 3RD edition, Oxford University Press. PP: 1035 – 1071.

Sarat, A. & Kearns, T.R. (Eds.), (1995) Scarry, E., Wald, P.M., Greenhouse, C.J., Hay, D., Weisberg, R., Sarat, A. & Kearns, T.R.; *Law's Violence,* University of Michigan Press, Michigan.

Save The Children, (2000) *Denied A Future? Volume 2, Western and Central Europe;* http://www.savethechildren.org.uk/scuk/jsp/resources/details.jsp?id=366&group=resources&section=publication&subsection=details.

Scraton, P. (1977) *Images of Deviance and the Politics of Assimilation: a Study of Labelling Processes and their Effect on a Special Education Project for Irish Gypsies*, B5d M.A., Liverpool, 27-5136.

Sexton, S.; Kinealy, C. (2002) *The Irish: A Photohistory,* Thames & Hudson, London.

Slater, R. (2005) 'My Terrifying 24 Hours as a Gypsy Gangster', *The Mail on Sunday,* Irish edition, 10/04/05, P: 36.

Smith, D (2002) 'The Hate Debate: Should Hate be Punished as a Crime?' The Institute for Jewish Policy Research, London.

Taggart, I. (2003) *Gypsy Travellers: A Policing Strategy – Why Don't You Just Move Them On?* London: HMSO, www.homeoffice.gov.uk/docs/gypsytravellers.pdf, (Accessed: 05/05/06).

Tatchell, P. (2001) 'Some people are more equal than others' pp: 54-71, in: P. Iganski, (ed.) Burney, E.; Jeness, V.; Lawrence, F.M.; Levin, J., Phillips, M.; Ray, L.; Smith, D (2002) *The Hate Debate: Should Hate be Punished as a Crime?* The Institute for Jewish Policy Research, London

Thomas, P.A., Campbell, S., *Housing Gypsies,* 2002 reissue by Rachel Morris, Traveller Law Research Unit, Cardiff Law School, www.cf.ac.uk/claws/tlru, (Accessed: 20/09/06).

Traveller Visibility Group, (2003) untitled website document: http://groups.msn.com/TravellerVisibilityGroup/judgeconolearysentencedtwotravellers.msnw, (Accessed: 07/05/06).

United Nations (2006) *World Drug Report* http://www.unodc.org/unodc/en/world_drug_report.html(Accessed: 20/02/07)

Whyte, G. (2002) 'Litigation and Travellers' Rights,' Social Inclusion in the Legal System, *Public Interest in Law in Ireland*, PP: 214 – 245,

Pub IPA.

Williams, L.P. (1994) *Gypsies and Travellers in the Criminal Justice System: The Forgotten Minority?* University of Cambridge Institute of Criminology, Cropwood Occasional Papers No. 23.

Williams, N. (2002) *Scoping London Diversity and Travelling People in London The Metropolitan Police Service Approach*, Diversity Directorate New Scotland Yard.

—. (2003) *The Customs of Travelling People: A guide to working with the Travelling Community,* unpublished paper, Diversity Directorate (DCC4), Metropolitan Police Service, London.

Wright, B., Duffy, D., Curtin, K., Linehan, S., Monks, S., Kennedy, H.G., (2006) 'Psychiatric Morbidity among Women Prisoners Newly Committed and Amongst Remanded and Sentenced Women in the Irish Prison System', *Irish Journal of Psychiatric Medicine,* 23(2): 47-53.

CHAPTER TWO

THE REIFIED "GYPSY": SYMBOL AND STEREOTYPE IN "THE TIME OF THE GYPSIES"

IAN HANCOCK

Time of the Gypsies, directed by Sarajevo-born film director Emir Kusturica is one of the very few screenplays with a Romani theme which utilizes Romani actors and the Romani language. In this respect it joins company only with Tony Gatlif's *Les Princes* and Robert Duvall's *Angelo My Love*. Despite this, it—predictably—relies heavily on stereotypes of mysticism and thievery, though their interpretation by the non-Romani audience differs from that of the Romanies who have seen the film.

In December 1985, Reuters released details of the arrest of a gang of Yugoslav kidnappers who, since 1980, had been stealing children from defenceless Romani families and selling them to Americans and Italians who were otherwise unable to adopt. The parents of the 100 kidnapped children, all Romanies, were too frightened to report these crimes. This was the subject of a report by Hans P. Rullmann which appeared in 1986 in *That's Yugoslavia* No. 5, and which was entitled 'Child slave trade in Yugoslavia: Gypsies (Romas) oppression.' The story also caught the attention of Emir Kusturica, a Yugoslav film producer known to Western audiences for his award-winning When Father Was Away on Business, which was released in the United States in 1985.

Kusturica, working with script writer Gordan Mihic, turned this human tragedy into a cinematographic fantasy, which he presented at the 1990 Cannes Film Festival as Time of the Gypsies, and which received a five-minute standing ovation.

The film itself is long, lasting two hours and twenty-two minutes, and is almost entirely in the Romani language, with English subtitles. It has been well received by almost everyone who has seen it; the Romani population of Cannes reportedly celebrated its release throughout the night until dawn; American Romanies who saw it struggled with the dialect, and were thankful that they lived in the USA. Given that its inspiration was the abduction of Gypsy children, and that the Romani experience it portrays is depressing, one is forced to question why it was so cheerfully received.

The story centers on Perhan, the son of a parentless household he shares with his demented uncle Merdzan, his grandmother Hatidza, and his disabled sister Danira. It opens onto a cold, wet, muddy day and a bride, still in her wedding gown, abandoned and angry because her new husband has passed out, drunk, in a hand-cart. Like a member of a Greek chorus, a forlorn character stands apart in this opening scene, and paints a word picture of Romani life. He is unnamed, and presented as a shaven-headed inmate from the Nazi camps where over a million Romanies were murdered during the Holocaust. Appearing periodically as a background character throughout the film, he speaks disjointedly of cruel medical experiments, and stresses that despite these things, he is still here. This seems to be a theme in the film; the pain and endurance of Romani existence. Another theme, which becomes apparent as the story progresses, is that survival rests upon adherence to core Romani values or Rromanipe; as Perhan moves further and further away from them, so do his misfortunes multiply. This is an allusion to the *kintala* or balance which must be maintained in Romani culture, and the *prikaza* or retribution the guardian spirits (the mule) mete out to those who upset it.

Perhan is a thinker and a dreamer; he has telekinesic skills, and can charm birds. He cares deeply for his sister Danira, and has vowed to pay her hospital bills in any way he can. He is in love with a young girl called Azra, who also lives in the village (which is not named in the film, but which is in fact Shutka, near Skopje). Although he wants to marry Azra, her mother is scornfully opposed to the union because, she says, Perhan has no money. She also comments on the fairness of her daughter's complexion, implying that Perhan is too dark for her, but at the same time deriding him for being the half-Romani bastard of a gadz.6 (non-Romani) soldier. Azra's mother has bought into the non-Romani value system, which places material wealth and physical appearance over moral or spiritual value; nevertheless Perhan is so smitten with Azra that her mother's words wound him deeply, and he attempts suicide. That he does this on Svuntogjorgi (The Feast of Saint George) is significant even to Muslim families such as Perhan's, because this slava (saint's day festival) is a time of family togetherness, and of ritual bathing in the river. Together with the singing that accompanies the flames of a thousand candles illuminating the dark water, this eerily beautiful scene is one of the most moving in the entire film. Perhan is in charge of operating the kiln in which chalk is burnt to produce the lime with which the houses are customarily whitened at this time of the year—and Azra's house is one of the whitest, despite the pollution of her mother's values. Luckily, Perhan's attempt to kill himself is foiled in time by a villager, who finds him hanging by the neck from a rope and cuts him down.

Although given a second chance at life, Perhan doesn't recover from his vow to become wealthy. He is obsessed with notions of becoming rich enough to pay for an operation on Danira's leg, and to win Azra's hand, especially when his uncle Merdzan boasted that he intends to seduce her, so when the crime lord Ahmed arrives, Perhan is easy prey.

Ahmed, originally from the village, had gone away to become rich through various criminal means, including selling children to Italians and forcing them to beg and steal for him. Perhan persuades Ahmed to hire him, and thereby begins his descent into criminality and his journey away from Rromanipe. Ahmed's superficial glamour, and his casual donations of cash to the villagers, make him appear as a hero, but as he wins their admiration and trust, he is at the same time undermining their spiritual and cultural integrity. Perhan especially was vulnerable to Ahmed's promises of easy wealth. He tells Perhan that if he and Danira go with him, he will be able to make the money so desperately needed for her operation. Soon after the three of them leave, Perhan begins to see that something is wrong because Ahmed is collecting young people, even babies, along the route. In a vision, Danira sees the muli'or spirit of their dead mother dressed in a wedding gown; her white dress, like the whitewash on the houses, symbolizes purity; but like the abandoned bride at the beginning of the film, and the materialistic, hostile mother in Azra's whitewashed house, and even like Ahmed's behaviour towards the villagers, the message is that what may appear to be good can in fact hide an evil which lurks behind it. But the appearance of her mother's muli, and the remembered words of her grandmother Hatidza that *'0 rat s'e tajna, o rat s'e zor'* (blood [i.e., family ties] is the secret, blood is the power), seems to comfort Danira in her nervousness. Her trepidation is justified nevertheless, because Ahmed leaves her alone at a hospital, and he and Perhan continue on into Italy without her.

Perhan is at first shocked to witness the human misery confronting him in Milan, but with time, he becomes inured and accepts it, becoming just as heartless a gangster as Ahmed in his overriding desire to acquire wealth. As his spirit hardens, so the environment which surrounds him becomes bleaker and more gloomy.

Eventually he returns home, but even before going to see his own family, he makes straight for Azra's house to demonstrate to her mother his new wealth and status. Here, he finds that Azra is pregnant with what he at first assumes to be another man's child. In his anguish, he seeks refuge in a tavern and tries to lose himself in drink. Instead, he finds himself confronting the decisions he has made in his life; he knows then that if he had not gone away and become dragged down into the criminal world of the gadze, things would have been very different. Still, he decides that he will not reject Azra, providing she gets rid of her child by selling it into a life of mendicity. She agrees, and together they set off for Italy, gathering up children along the way just as Ahmed had done, though this time to be put to work for him.

Back in Milan, Azra goes off by herself to have her baby, but doesn't return. Then Perhan finally goes to look for her, he discovers the baby alive but Azra dead, having died giving birth. His misfortunes are compounded by a violent storm which washes away his cache of gold. Now without the woman he loves, and with his savings all gone, Perhan believes himself to have sunk as low as possible, but finds 'in the lowest deep a deeper still,' when he learns that Ahmed, whom he'd looked upon as his mentor, had deceived him. Instead of taking care of Danira's operation, he had forced her into begging for him, capitalizing on her deformed leg as a money making source of pity.

Perhan leaves at once for Rome, where Danira is. She receives him without recrimination. and in fact they seem closer than ever before; the ties of blood overcome everything else. Perhan puts Danira and his new baby (also named Perhan) on a train back to Yugoslavia. promising to join them in due course. Meanwhile, consumed with the desire for revenge, he goes to Ahmed's wedding and murders him using his telekinetic skills: he makes a fork fly across the table and fix itself in Ahmed's heart. It is at this point that Ahmed realizes how overriding are the bonds of blood and family loyalty. In an ending which surprised the non-Romani audience but not the Romanies who saw the film, Perhan too died in this scene, when Ahmed's bride then killed him in retaliation.

The penultimate scene is of Perhan's funeral. His corpse is lying in an open coffin with gold coins on its eyelids, while his son, now looking like a miniature version of his gangster father, is reaching in to steal them. In the very last scene, little Perhan is running off with the coins, chased by Merdzan who saw him do it. Reaching the crossroads, Merdzan is faced with 'a choice: in one direction lies the road down which little Perhan is fleeing, while the other leads away from him, to the church. This is the one Merdzan chooses. It has taken the examples of his nephew Perhan, and now Perhan's son, to make him realize the futility of his own errant ways, and so he turns toward God to redeem himself before he ends up like Perhan and Ahmed.

The film is heavy on imagery; it begins and ends with a wedding, and the spirit of a bride, Perhan and Danira's mother, also appears halfway through. In Rromanipe, marriage is the fulcrum, and it is difficult to believe that Kusturica didn't know this when producing the film, or else—and what is more likely—that the members of the almost entirely all-Romani cast hadn't helped him to understand it. The *kintala* or balance one must maintain involves two sides equally weighted; stray from *Rromanipe* and the balance shifts. The two sides are just one manifestation of the overall duality which characterizes the Romani world view: Rom and gadjo. God and the devil, purity and defilement, luck and misfortune, male and female, child and adult. Marriage is the boundary between one's status as a boy or girl (in Romani, *chav* or *chej*) and man or woman (*Rrom* or *Rromni*). This boundary has nothing to do with biological age; a girl who marries at age twelve becomes a Rromni or woman, because the word is synonymous with 'wife,' and a boy becomes a husband and a man. In Rromanipe, marriage and parenthood are seen as healthy and part of the natural order, while single status is seen as bad. It can bring retribution (*prikaza*) from the guardian spirits (the mule) in the form of bad luck (*bibaxt*) or illness (*nasvalipe*). The non-Romani world is an evil place fraught with danger, because those who inhabit it are seen, from the Romani perspective, to have no moral or spiritual structure. Perhan's entry into that world meant an automatic rejection of his own, and seen from that perspective his ever-increasing misfortune, and ultimately his death, are justified.

Time of the Gypsies meant quite different things to the two audiences—Romani and non-Romani—who saw it. For the Romani audience, it was a cautionary tale replete with symbolism which had immediate relevance for them. For the non-Romani audience, the same message was apparent. But without a knowledge of the Romani world view, much of the symbolism must have been interpreted simply as a part of the 'mystery' of the Romanies. This was reinforced by the quite unnecessary introduction by Kusturica of magical tricks, moving cans, flying forks, floating scarves and women. These things are not part of real Romani life, but they are certainly what caught the imagination of the reviewers, and what those reviewers thought belonged in a film about Romanies; *Variety*'s account of the movie (May 24, 1989) called 'the film's best moments ... its flights into fancy.'

Without a proper understanding of the Romani world of southern Yugoslavia, a non-Romani audience must react on one or two levels only;

the references by its characters to 'Gypsy whore,' 'lying Gypsy bastard,' and so on would not seem out of place to an audience (especially a European audience) who probably use such expressions themselves. But the contemporary Romani world and this kind of social behaviour cannot be understood without putting it into the context of history; the centuries of dehumanizing slavery during which such epithets were flung at them by their owners have resulted in the internalization of that disdain, but the audience won't learn this from the story. Nothing was apparent in the film which told the audience why characters such as Ahmed and Perhan were drawn into crime, even crimes against their own people; nothing was explained to the audience about why the Romani village was so squalid. Those watching the film knowing little about Romanies beyond the fictional stereotype (and this includes most of the American audience) are likely to come away from the theatre with the unquestioning assumption that that's just how Romanies are. In their incredible journey by foot and on horseback from India into Europe a thousand years ago, the ancestors of the Romanies remained together, one people speaking one language. Only after encountering the European population did they become fragmented and thereafter subjected to centuries of oppression and infamy, enslavement, and genocide.

It might be argued that artistic licence needs no defence, and that similar portrayals exist for all kinds of populations. This is certainly true; but while the public is entertained with fictional representations of, for example, Italian criminality in the countless films and novels dealing with the Mafia, it is also made aware in the school curriculum of the great contributions Italians have made to society: Columbus, Botticelli, and Michelangelo are names known to everyone. This is not so in the Romani case. Two recent books with Romani themes are Julian Thompson's *Gypsyworld* which, written for young adults, deals with 'middle-aged Gypsies' who drug and kidnap children. The other is Joe Gores's *22 Cadillacs*, on the cover of which is written "Gyppoes: Con artists, scammers, liars, thieves and dangerous charmers, Gypsy Americans are one nation united in street crime." Films such as *King of the Gypsies* (now in re-release at Blockbuster Video), *Golden Earrings*, *Hot Blood*, television documentaries such as *Geraldo*, *America's Most Wanted*, and *Now It Can Be Told* consistently and unrelentingly present a negative and unrealistic stereotype of Romanies[1].

---

[1] Since writing this *Daddy Daycare* has been released by Columbia Pictures (Spring 2003), written by Geoff Rodkey and produced by Matt Berenson. Aimed specifically at children and starring Eddie Murphy, it includes a character called the "Old Gypsy Woman" who runs a child-care centre. The children are kept

Unless children also learn in school about the Romani contributions to the world, to the arts and to music, unless they learn that such well-known personalities as Django Reinhardt, or Carmen Amaya, or Carlos Montoya, or Lafcadio Hearne, or Bob Hoskins, or Charlie Chaplin[2] are of Romani descent, they will continue to get an unrelievedly one-sided picture of a people only ever represented as immoral and dishonest.' A balanced view, the non-Romani public's *kintala*, will never exist for them.

Since such behaviour is not genetically determined, but characterizes only individuals defined by social behaviour, the notion that Romanies are a behaviorally defined population is simply reinforced, allowing those such as Gores or Geraldo Rivera to make hurtful and damaging statements about Romanies that they would never make about other racial or ethnic

---

locked in an underground bunker, and she appears dressed in the requisite bandanna and hoop earrings, to the accompaniment of Gypsy violins.

[2] Chaplin's Romani ancestry is not once mentioned in Richard Attenborough's biographical film of the same name; in fact, Chaplin is presented in that film as being of Jewish descent, which is commonly believed, 'despite the contradictory evidence prompting it, particularly the evidence coming from Chaplin himself [in his own autobiography]' (John McCabe, in his Charlie Chaplin. p. 2). David Robinson (Chaplin: His Life and Art, p. 2) discusses Spencer Chaplin's marriage to a seventeen year old [G]ypsy girl called Ellen Elizabeth Smith, in the parish church of St. Margaret, Ipswich. The witnesses to the marriage included the girl's father, who was illiterate, but no Chaplins, so perhaps the family was not happy with the match. Ellen Elizabeth died in 1873, at thirty-five, and no photograph of her survives, so we can only speculate that it might have been she who originated the striking looks, jet hair and fine eyes that became the Chaplin heritage.

The widespread misassumption of Chaplin's ancestry is a curious one, and serves to illustrate how the lack of information about the Romani people can lead to such popular myths. The notion that Chaplin was Jewish can probably be traced to his role in The Little Dictator, an anti-Nazi film in which he plays the part of a Hitler look-alike. The film, now half a century old, was thought naturally enough to have been stimulated by Jewish concern, as indeed it probably was—but so little was known then, as even now, of the immensity of the Nazi campaign against Chaplin's own people, that it was automatically assumed that he was Jewish. Even today, the identity of the Romani victims of German skinhead attacks and the deportations organized by the German government is hidden behind the phrase 'Romanian refugees'; similarly the fact that hundreds of thousands of Romanies being tortured and imprisoned in what used to be Yugoslavia, and are being used to test the minefields in Banja Luka and elsewhere, is kept from the American public because their identity is concealed behind the names 'Muslims', 'Serbs', or 'Croats', despite their being not only a culturally and linguistically, but also a genetically distinct population.

populations. Kusturica was to some extent also guilty of this. By his own admission, he didn't want Time of the Gypsies to be 'just a realistic movie ... if you're going to make a structure based on Gypsy life, you have to change your form.' It is a great pity that, in exploiting both the present-day tragedy of Romani life in Europe, and the mystical, magic Romani image created in the European mind, he did not also take the opportunity to educate the audience about the real history of the Romanies.

In an earlier review of Raggedy Rawnie, another 'Gypsy' film (Traveller Education, 23:17-23, 1990), I criticized what disturbs me with the production under review here: the exploitation of the mystical at the expense—even to the exclusion—of the actual. Ironically the one Romani film, Les Princes (produced by Tony Gatlif, himself a Romani) which does deal with the depressing realities of Romani slum life in the gitaneries outside Paris, was neither well promoted nor positively reviewed. One can only conclude that the need on the part of the general population to be fed Gypsy fantasy is tenacious indeed, and goes beyond a desire for mere entertainment. But until the Romani people are demystified, credit for their real contributions and accomplishments will continue to be attributed to others, and public outrage at the contemporary realities of anti-Gypsyism will remained muted.

Acknowledgements

An earlier version of this essay appeared in the *Journal of Mediterranean Studies*, 1997 Vol. 7, No. 1: 52-57. (ISSN: 1016-3476)

References

Debord, G. (1994) *The Society of the Spectacle*; New York: Zone Books

Fanon, F. (1967) *Black Skin, White Masks*; New York: Grove Press

Kaplan, A. (1993) *Women and Film: Both Sides of the Camera*; London: Methuen

Mirzoeff, N. (ed.) (2002) *The Visual Culture Reader* (2nd ed.) London: Routledge

May, J. *Nomadic Identities: The Performance of Citizenship;* Minneapolis: University of Minnesota Press

Easthope, A. (1993) *Contemporary Film Theory*; New York: Longman

Nowell-Smith G. (ed.) (1996) *The Oxford History of World Cinema*; Oxford University Press

Turner, G. in J. Fiske (ed.) (1988) *Film as Social Practice*; London: Routledge

CHAPTER THREE

TRAVELLING STYLE: SOME ASPECTS OF JOHNNY KEENAN'S PIPE-PLAYING

MÍCHEÁL Ó HAODHA AND DAVID TUOHY, UNIVERSITY OF LIMERICK

> In the first rushing, I thought it was Doran who was playing! The style and 'dialect' are very close to him and this was a reel greatly favoured by Johnny himself. The version and harmony accompaniment are practically identical with what he used to have, but Keenan's personal style is noticeable throughout, especially in the tight fingering. As with Johnny, he is gentle and skilled in his harmony accompaniment. I am familiar with this reel as 'The Mountain Lark' (O'Neill No. 1244). "McLeod's Reel" is well known, but here Paddy's piping is worth hearing and having to hand.
> —Slip Jig: *Drops of Brandy*

> …he told me he was making keys with silver spoons…and he was making reeds out of bits of cane…he used to do all his own work with his pipes, anything would go wrong with them he was able to do them.
> —Mrs. M Cash, a close relative of Johnny Doran – *The Long Note* (1988)[1]

In this essay we examine some aspects of Paddy Keenan's musical style and discuss some aspects of his music that are often considered typical of what is sometimes termed the "Traveller style" of Irish pipe music. We also outline a number of ways in which Paddy Keenan has contributed to the development and of certain technical facets of the uilleann pipes as they continue to evolve within the category of music known worldwide as traditional Irish music. Generally acknowledged as the most accomplished uilleann piper performing today, Paddy Keenan is certainly one of the most brilliant musicians of his generation, a generation which has seen traditional Irish music scale unprecedented heights in

[1] The Long Note was a series of music programmes on RTE Radio 1 (Irish radio), which were broadcast in 1988.

terms of global appeal and recognition. Paddy Keenan is also unique in that he has inherited an enduring musical legacy from a minority within Irish culture – the Irish Travellers. In the pantheon of Traveller musicians he can rightfully claim his place alongside such piping legends as fellow Travelling musicians John Cash and Johnny Doran. Paddy Keenan was born in Trim, County Meath, in the heart of the Irish midlands. His father, John Keenan Sr., from neighbouring County Westmeath was a very well-known musician in his day as was his father before him. His mother was from County Cavan, an area closer to the border with Northern Ireland. Her maiden name was Mary Bravender. Paddy was about ten years of age when he first began playing the uilleann pipes [2]and four years later he had his first major public performance when he took part in a concert at Dublin's famed Gaiety Theatre. He later played with the rest of his family in a group called *The Pavees*.[3] At seventeen, Paddy left the traditional music scene and Ireland behind and travelled around England and throughout Europe where he concentrated on playing blues and rock with a number of bands. He then returned to Ireland after a couple of years and began playing music with two well-known traditional singers and musicians Tríona Ní Dhomhnaill and her brother Mícheál Ó Domhnaill. Dublin fiddler Paddy Glackin, who was to join Paddy Keenan as a founding member of the world-reknowned Bothy Band at a later date, joined the this trio who later became five with when they were joined by flute player Matt Molloy. Within a short space of time the group was augmented by the arrival of two other seminal influences in the Irish traditional music "revival" of the 1970s - accordion player Tony MacMahon and guitarist/bouzouki player Donal Lunny. These various musicians combined to form the group "Seachtar"[4] Seachtar played a number of headline events but, like many an Irish band before them there was a certain fluidity to the band's personnel from its very beginnings. Circumstances soon forced Tony MacMahon to drop out and when the rest of the band decided to turn professional Paddy Glackin took his leave and was replaced by Donegal fiddler Tommy Peoples - who was subsequently replaced by fiddler Kevin Burke. The band decided on a name change and when Mícheál Ó Dhomhnaill came across a Scottish photograph from the 1890s of a group of tattered Irish migrants who were also musicians – "The Bothy Band" was born! The "Bothy" was a reference to those migratory Irish laborers who worked in Scotland and throughout the rest

---

[2] Uilleann Pipes – (Lit: "Elbow Pipes" – the word Uilleann being the Irish word for "elbow").

[3] Pavee is a Traveller Cant/Gammon word meaning "Traveller" or "pedlar".

[4] The word "Seachtar" means *Seven* in Irish Gaelic

of Britainin the last century and were then housed in stone huts known as "bothies." The Bothy Band became one of the most influential bands of the 1970s and are considered to have changed the face of Irish traditional music as its revival strengthened towards the end of the Millenium. The Bothy Band were seminal in "modernising" Irish traditional music through the amalgamation of a driving rhythm section with traditional Irish tunes in new and innovative manner that had never been heard prior to this. To see the band playing live was considered quite an event, an experience which one music reviewer referred to as quite similar to "being in a jet when it suddenly whipped into full throttle along the runway.!"[5]

Central to the special and explosive charge unleashed by The Bothy Band was Paddy Keenan, one of the band's founding members. His virtuosity and the ferocious energy with which he played the uilleann pipes on The Bothy Band's albums meant that he was one of the band's driving forces, a fact which led fellow band-mate Donal Lunny to describe Paddy as "the Jimi Hendrix of the pipes!". Since the heady days of The Bothy Band, Paddy's musical style has developed and his innovations in uilleann pipe music continue to enhance a wide range of traditional Irish "numbers" with a new vibrancy while remaining true to the older, core rhythms of the tunes. One area in which Keenan has contributed more than most to the "revival" and development of the Irish uilleann piping tradition relates to the construction and enhancement of certain technical innovations as relating to the instrument itself. In this aspect Keenan is drawing on a very old and as yet little-acknowledged aspect of Irish traditional music whereby Traveller musicians such as the Cashes, Dorans, Dohertys, Dunnes and Keenans to name but a few, brought their skills as Travelling craftsmen to the service of the tradition. That Traveller musicians, in particular, made a substantial contribution to the development of many technical aspects of instrument-making is a fact that surfaces consistently in the margins of the few essays and books that have been written to date, where the influence of the Traveller tradition on Irish music is discussed. Traveller John Doherty, who is probably the most famous fiddler ever to have come from County Donegal, a county which has specialized in producing talented fiddlers from one generation to the next in addition to being a consummate tinsmith, was also known as a good maker of tin fiddles. These which were ideal for beginners who did not wish to spend a lot of money on the purchase of a new instrument but who nevertheless wished to find out time whether the fiddle suited them as an instrument, as indicated in the following account by "Pecker" Dunne:

---

[5] Cited in *About Paddy* - http://www.paddykeenan.com/about.htm

> There aren't too many musicians today who could make their own instruments. They are all factory-produced and made to certain specifications. However, that wasn't the way with the older Travelling musicians. I could make a fiddle even today if I had to. Many is the time that my father and I made our own fiddles when we were young. It was a skill that all the good Traveller musicians had. We were able to make anything else that could supplement our income. You name it and the Traveller musician could make it when needs must. Horses reins and harnesses, fishing rods, tin cans, filters, poteen stills, uilleann pipes, fiddles and whistles. They were all the same to us. There was a great fiddler in Donegal called Johnny Doherty who used to travel that county in the 1930s and 1940s. He was able to make fiddles from tin which were ideal for beginners who were interested in learning the fiddle but were not sure whether they would be able to make a "fist of it". The Romany Gypsy musicians in England called the Boswells made their own fiddles as well. They made wooden box fiddles that were a similar style to the Dunne fiddles. Sometimes you would get an old fiddle that was broken up and take the neck out of it. And then you'd make the square box part of the fiddle and mold the two together. I often made them when I was young and my uncle Briany Dunne was a great man to make a fiddle (Cited in Dunne and Ó hAodha 2004, 47).

Patrick "Pecker" Dunne, another renowned Irish Traveller musician, who played with such luminaries of the Irish "Folk Revival" as the Dubliners and balladeer Margaret Barry has also acknowledged the contribution of Travellers such as his own family to the instrument-making process:

> My father broke his bosh one night when he was in Waterford. He was after having an argument with my mother. He had a few drinks on him and he got mad and danced on the fiddle…Anyway she took up the broken bits of the fiddle from the floor of the pub and brought them home. And the next morning my father and I got up and went down the town to find a few bits of timber. We found these boards that people used to call stuff boards. They were the boards that would have had cloth wrapped around them. They were great pieces of timber with which to make a box fiddle because they were well-seasoned. We brought home three or four of these boards and set to work on making a new fiddle. The next day we headed into Waterford again, my father with his new box fiddle under his arm. We played the town the same day….One time I was travelling in County Cavan myself and I broke the strings on my banjo. I had no spare strings, nor did I have the price of any new ones. So I went into a bicycle shop and asked the owner of the shop for some cable wire. This was the cable wire that they used to change the bicycle gears with years ago. He gave me a new pack of the stuff, fair play to him and I opened it up and made 4

> strings out of it. It didn't sound anything as good as a real set of strings but it did the job. I played away on them until they got me the price of a new set of strings. "When needs must", as they say. (Dunne 2004, 47-48)

That this tradition of continuous innovation and development in terms of the construction of the musical instrument itself is clear from the big contribution which Paddy Keenan has made to development of the uilleann pipes. While every uilleann piper would probably say that their instrument is unique and particular to themselves as a musician there is no doubt that Paddy Keenan has contributed hugely to the technical enhancement of the uilleann pipes both in terms of its physical construction and the texture and range of sound that can now be achieved with the instrument. The body of Paddy Keenan's uilleann pipes was found by Keenan's father, John, in a second-hand shop on Capel Street Dublin many years ago. It is made up of three drones and regulators which are constructed of blackwood and ivory and silver mounts. The chanter was made by famous Dublin pipemaker and musician Leo Rowsome in the early 1930s. This is made from ebony and the wood is said to have come from the west coast of India. It is said that there is an actual historic link between this chanter and the Doran family, a Traveller family who are acknowledged as seminal influences not only on the Traveller piping tradition but on the Irish uilleann piping tradition as a whole. The oral tradition has it that the chanter originally belonged to one of the Dorans who were the first to begin a process of filing and under-cutting of the chanter, a process which has continued and evolved after the chanter passed into the Keenan family. Interestingly, the pipes' bellows were actually made by the Keenans, Paddy's father John Sr. constructing the bellows and the bag being made by Paddy himself. Although he belongs to the current generation of Traveller uilleann pipers in an era when musical instruments are frequently "factory-produced" or made by machine, Paddy himself has made an important contribution to the construction of the bag. Traditionally the bag which accompanies the uilleann pipes was made out of basil although Traveller musicians who were always versatile often used a range of other materials when they were available, if they give a better sound or when the products they preferred were scarce. For example Paddy's father was not unusual in creating a bag out of an old piece of rubber such a as the tube of a lorry or a trailer. Vinyl was a product which increasingly came to be used for the construction of the bag from the 1970s onwards, in particular a type of vinyl known as leatherette and Paddy Keenan was one of the first pipers to use this new product. Vinyl has brought a new elasticity to the bag and helped regulate the air pressure which the piper forces out of the bag while playing. Greater elasticity has helped further develop the quality of sound and the nature of the notes that can be produced with the bag and the uilleann pipes as the rate at which the piper is able to force out the air pressure greatly affects the tone which the chanter produces. The bag itself as constructed by originally by the Keenans is also a new departure in terms of uilleann pipe design. In the past a single bag was prone to splitting or bursting under the huge pressure which is applied by the piper as his elbows work the bellows. The Keenans came up with the idea of creating a bag inside a bag – a double-bag effect – whereby the vinyl seams are tougher and more durable under pressure. The older and more traditional skills of the "smith" have also come into play as regards the manner in which the bag is attached to the rest of the instrument. While in the past craftsmen tended to use waxed hemp to attach the main stock and the blow pipe to the bag, the Keenans and Paddy in particular have modified this process by the actual soldering of a small plate onto the back of the main stock and the fitting of a screw attachment there. This has ensured an enhanced durability in the instrument whereby it is virtually impossible to burst the bag irrespective of how much pressure is applied to it during the playing process. This is an important innovation in relation to the construction of the uilleann pipes as in the past it was the tradition to ask a tailor or somebody with a "good eye" to stitch the bag by hand. This stitching processs, while an artform in itself also meant that the bag of the uilleann pipes was left vulnerable to leaks as it involved the piercing of the actual leatherette or vinyl. Today the seams of many of the bags are glued to the instrument as opposed to being stitched, a fact which renders the potential for leakage negligible. Another innovation which has had a significant effect on the widening sound-range and tone of the notes that can be produced with the pipes, the adoption of which has been driven primarily by Paddy Keenan applies to the reed of the uilleann pipes and in particular, the construction of the reed. This aspect of the instrument's evolution almost certainly owes a significant debt to those generations of Keenan family pipers who went before Paddy, in addition to Paddy Keenan himself. While the reed is one of the smaller physical attributes of the uilleann pipes, it is nevertheless one of the most vital as the standard of the reed has a huge influence on the quality of the sound which a particular piper is capable of producing. It is also more often than not the exemplar of a piper's individuality as an artist and the source of his "style", or that particular sound that is recognisably his/hers. In the case of Paddy Keenan, the stories regarding his propensity for experimentation as pertaining to the construction of the reed have nearly achieved "mythic" status in musical circles. Like many Traveller musicians before him, including his own father John, Paddy Keenan has experimented with the shape, texture and craftsmanship of the reed from his earliest days playing the uilleann pipes.

It is said that his father was actually teaching him to fashion reeds of different shapes even prior to his first lessons on the pipes. Keenan's propensity of experimentation was such that he often received "strange" looks from his neighbours when he hung his range of freshly lacquered reeds from the branches of a tree in the family's back yard. A description of the intricacies of his reed-making process would form the basis of another paper but a few outline aspects of the process can be noted. Keenan generally uses pieces of cane for his reeds and has a certain preference for Californian cane which – as he lives mainly in the U.S. – he has access to reasonably easily. He sands the cane down using a circular object, usually a glass bottle, wrapped in sandpaper. This process is enhanced by the use of sandpaper and the Keenan shapes the blades of his reeds by "eye" only as opposed to using measuring devices such as mircrometers as used by other reedmakers. When the blades of the reed are shaped to his satisfaction he ties them together and inserts a staple made from a metal tube into the reed. He then binds this staple to the tube with shoemaker's thread which is airtight and very strong. Further sanding and shaping takes place and the reed is regularly tested for its sound – a process known by pipers as "crowing the reed". Once the desired "crow" or pitch of sound is reached a thin belt of copper known as a bridle is fitted to the reed. Keenan tends to go for flexibility and elasticity in all of these technical materials so that when he blows his different notes – it is possible for him to achieve different sound levels, sound pitches and textures of sound. Elasticity in the construction of both the reed and the bag is a vital element in Keenan's approach to both uilleann pipe construction and playing with the consequence that many other pipers who "give it a go" with his pipes say that they find it very difficult to play his chanter in tune or get the same sound from it that Keenan does.

Many aficionados of Irish music and the uilleann pipes in particular state that Keenan's musical style – often referred to by the term – "flowing" can be attributed to the his own ethnic heritage and the preference of a number of Traveller piping greats such as Johnny Doran for the open-fingered style, a style which is often said to be a hallmark of the Traveller piping style, one incorporating the use of flowing sequences of triplets. Seoirse Bodley, a noted composer and one-time Professor of Music at University College Dublin summarised the "open" Travelling style with particular reference to the piping of Johnny Doran as follows:

> ...in the reels you have this...flowing series of triplets that just flow up and down in the easiest way and I think it's the most striking aspect of his style in many ways (The Long Note, 1988)

A closer examination of Johnny Keenan's piping technique however reveals that while his style is indeed predominantly an open one – incorporating a particular fondness for intricate ornamentation and the use of flowing segments of legato triplets being the style "marker" - his style also incorporates significant elements which cannot be defined as open at all, including the use of staccato triplets and "new" triplets which lie in the "ambiguous" area is between both the legato and staccato styles.

While the open-fingered style is today considered most "typical" of the Traveller style of piping, it is arguable that this is only so because the great Johnny Doran, who was to have such a huge influence on succeeding generations of Irish pipers both Traveller and "settled" chose to play predominantly in this style. As there are so few recordings extant of Traveller pipers[6] it is very difficult to say with any certainty whether the open-fingered style incorporating the use of (predominantly) legato triplets was ever really the most typical style of all Irish Traveller pipers. Although often compared to Doran, it is worth noting that Paddy Keenan was already in his early twenties by the time he was to hear a tape of Doran's playing. Consequently, Keenan's musical dialect or style was already at a very well-formed stage in piping terms before he had even had the opportunity to hear Doran's sound for the first time. It is also almost certain that Paddy Keenan has had access not only to a much wider and diverse range of musicial influences – e.g. jazz, rock, country-western – than Johnny Doran ever did, but also to a wider range of pipers and piping styles. These include the acknowledged influence of "staccato" piper Liam Walshe and legendary New York-born piper Patsy Touhey, both of whom were musicians from the settled community and both of whom had a preference for a more staccato style of piping as opposed to an entirely open-based style. Ironically, it may be that those people who claim that a direct "line" between Johnny Doran's "Traveller style" of piping and Paddy Keenan's may be correct in their assumptions but not necessarily for the reasons that are most commonly claimed. In fact Doran may have been a strong catalyst for innovation in uilleann pipe piping and as applied to the playing of triplets in particular, innovations which have been carried

[6] Incredible as it may seem there is in existence only one recording of Johnny Doran who is widely acknowledged to have been one of the greatest Irish pipers ever, a recording which was completed very shortly before his death (RIP) in 1950.

a step further in recent decades by Paddy Keenan. Attentive listening to the uilleann-pipe techniques of both Doran and Keenan reveal the existence of a range of "new" or "different" notes which can be defined as neither staccato nor legato. While these notes are clearly articulated they do not identifiably belong to either category, a fact which has resulted in these notes being termed "defined triplets" by some musicians and musical theorists (Ring 1992, 55). These triplets are played by using a type of close fingering but in a manner which is still less forceful or "defined" than that which is the norm during staccato fingering. Conversations with Traveller relatives of Johnny Doran and other pipers suggest that this was style of triplet was also a regular, if relatively unacknowledged aspect of Johhny Doran's musical technique also although most people would argue that Paddy Keenan has brought non-legato/non-staccato style to a new level in terms of sound and development. There are many musicans who would argue that the repetition (often at incredible speed) of successions of triplets – whether legato, staccato or "defined" - is another hallmark of the Traveller piping style. Whether this is true or whether this categorisation of the Traveller style is another "stereotype" which has assumed the status of "truth" is a source of debate which would need a much longer essay than our current one to tease out. What is beyond doubt however is that Paddy Keenan has set new "markers" in terms of triplet playing for uilleann pipers through his range and mastery of the playing of sequences of triplets. In fact, his triplet playing on some traditional Irish music tunes is so exceptional that it has arguably become a showcase for his own absolute mastery of the instrument. A close analogy – albeit as related to a completely different instrument and musical genre – would be with the absolute control of his musical instrument expressed by renowned trumpet-player Miles Davis on "defining" albums such as Kind of Blue, Bitches' Brew and Sketches of Spain. Keenan's total mastery of the uilleann pipes comes to fruition in another aspect of his musical style for which other pipers frequently express their admiration and (sometimes) envy! It is an aspect of his style which we can only briefly allude to within the framework of this essay but which is so characteristic of Keenan's "sound" that no discussion of his piping could not fail to mention it. This technique known as "backstitching" involves the rhythmic repetition – albeit that each repetition is accompanied by different and more intricate forms of the same notes – of the particular triplets as a tune progresses sometimes using a particular note as a primary reference point – the note C# is a particular favourite of Keenan's. "Backstitching"[7] is a term used

[7] See Mitchell and Small (1986: 25) for another explanation of the technical term

initially by and in relation to some of the older Irish pipers who emigrated to the U.S. at the turn of the century – hugely influential musicians such as Patsy Touhey – who would take a particular triplet – e.g. G, F# and E and continue referring to it[8] throughout the progression of a tune while using one particular note as their primary reference point or "base" – so as to avoid getting "confused" or "lost in any way. For those readers who wish to investigate this aspect of Keenan's style further it is worth listening to the manner in which he plays a well-known Irish tune such as "Harvest Home" on the various albums he has produced to date. On this tune and others Keenan frequently uses the "backstitching" to maximum effect while using C# as his reference point, thereby utilising a musical "turn" which gives him the freedom to demonstrate his virtuosity, speed and explosive power in terms of triplet playing and improvisation.

Conclusion

In this essay we have examined some aspects of Paddy Keenan's musical style and technique which act to distinguish him as an Irish musician and as a musician drawing on the musical well that is the Traveller piping tradition. We have outlined a number of ways in which Keenan has contributed to the development of certain technical facets of the uilleann pipes in the modern era. We have also discussed some aspects of his music that are often considered typical of what is sometimes termed the "Traveller style" of Irish pipe music and pointed to some of the difficulties inherent in definitely categorising or "boxing in" the particular style of a musician who has absorbed a very diverse range of influences, including traditional Irish music styles as developed by both Traveller and non-Traveller musicians. While it can be argued that Keenan's playing technique as outlined in the use of legato and "defined" triplets has been strongly influenced by the Traveller style, and Johnny Doran in particular, it is also equally true that other aspects of his style, his virtuosity on staccato notes being a prominent example is as much indebted to the non-Traveller Irish musical tradition.

Keenan has continued to develop and mature his style in the years since the break-up of the The Bothy Band as he has pursued a solo career. He continues to record and to play at many of the major annual festivals throughout the U.S., Canada, Britain and Ireland and more local-based sessions or ceili's (dances). One of his most recent albums, "Na Keen

---

"backstitching".

[8] Touhey would continue referring to the note in an ascending run.

Affair," is available on the Hot Conya Records label and details of his tours and ongoing musical projects can be found at: http://www.paddykeenan.com/

Acknowledgements

The authors of this essay would like to thank the many musicians (both Traveller and non-Traveller) who generously shared their insights into the pipe music of both Paddy Keenan and Johnny Doran for the purposes of this essay. A special word of thanks is extended to Paddy Keenan and the members of the Doran and Rooney Traveller families in England for the generosity of their help in the background research for this essay.

References

Breathnach, B. (1977) *Folk Music and Dances of Ireland*; Cork: Mercier Press

Buckley, A. (1979) *Considerations in a Stylistic Analysis of Uilleann Pipes*; Stockholm: Musikhhistiriska Museet

Garvin, W. (1978) *The Irish Bagpipes: Their Construction and Maintenance*; Belfast: Blackstaff

Mitchell, P. and Small, J. (1986) *The Piping of Patsy Touhey*; Dublin: Na Píobairí Uilleann

O'Canainn, T. (1978) *Traditional Music of Ireland*; London: Routledge and Kegan Paul

O'Neill, F. (1987) *Irish Minstrels and Musicians*; Cork and Dublin: Mercier Press

Ring, B. (1992) *Travelling style : the Uilleann piping of Paddy Keenan;* (M.A. Thesis), Cork: UCC

CHAPTER FOUR

IRISH TRAVELLERS, NATIONAL SCHOOL AND THE EDUCATIONAL EXPERIENCE

MÍCHEÁL Ó HAODHA

This essay gives a brief overview of a unique and exciting piece of research as undertaken by a group of Traveller women from Limerick city, in south-west of Ireland, the findings of which were published in a book entitled – *Whiddin to the Gauras – Talking to Our Own* (2005). Most important of all however is the fact that it gives a small glimpse into what it is like to be considered "different", "other" or labeled as a "problem" in modern Ireland. There is widespread acknowledgement amongst both the Travelling and settled communities that education is a crucial asset in order for young people to live their lives with dignity in present-day Ireland. Achieving a certain level of education is necessary to accomplish many of the most mundane and tasks that each citizen of the State undertakes on a daily basis – e.g. taking the correct exit onto the motorway; passing the exams to be gain a haulier's licence or negotiating the numerous roundabouts which seem to have mushroomed on the approach to every Irish town since the days of the "Celtic Tiger" economic boom! Having appropriate access to an education also aids a person's development in terms of their identity and the ability that each individual has to affirm or negate resistance to that identity. This is particularly true of individuals who are members of a small minority ethnic group such as the Irish Travellers. As a group who have been primarily a non-literate community in recent generations educational access is a more pressing issue for this minority than for many others in present-day Ireland. Although more and more Traveller children now complete primary school, very few go on to secondary education and only a handful complete third level. This means, among other things, that policies affecting Travellers, and the research on which they are based, are framed and implemented by outsiders, policy-makers who are frequently not familiar with Traveller culture and way of life. While it has been the case that the Traveller community have been researched in a variety of ways over recent years, it has generally been the case that outside "experts" have been commissioned to analyse or explore a particular aspect of Traveller culture or life. While this research has sometimes had a collaborative element and involved the Traveller community in its research process, it has more often been the case that there has been little or no consultation with the Traveller community. This project was an attempt to redress this balance. Four Traveller women from Limerick—Mags Casey, Bridgie O'Donoghue, Ann O'Donoghue and Ann O'Driscoll—trained as qualitative researchers with project leader Eleanor Gormally, a lecturer in Education from Mary Immaculate College. The research, which was designed and carried out in a collaborative manner, focused on interviews with a small group of Limerick Traveller children about their lives, and especially school. As is customary with many community development projects, the manner in which the project was carried out was accorded just as much importance as the project outcomes. This is reflected in the structure of the book, with the first two out of three sections dealing with the women's/Traveller parents reflections on their own educational experiences and on the methods they used while conducting their research. The third section discusses the content and implications of what the children said during the interviews. The book charts the ongoing process that were the training and intercultural dynamics of the project and makes clear that working with minorities such as Travellers in a collaborative manner – where the members of the minority are the subjects as opposed to the objects of the research – can be a very productive and successful research process. The book gives an overview of the complex relationship that exists between Traveller parents, Traveller children and the Irish educational process, and outlines the extent of Traveller educational inequality within the Irish educational system today. The second half of this qualitative study makes for particularly interesting reading as it gives a very clear insight into where Traveller parents and children have concerns as regards their interaction with the educational system. These concerns are cumulative and are a consequence of the experiences of a number of generations of Travellers with an Irish society, the general thrust of which has been assimilative and generally ignorant of most aspects of Traveller identity and culture. Parental involvement is undoubtedly a key element in the education of children and the establishment of a successful and ongoing relationship between home and school. While there are exceptions, this project shows that for the majority of Traveller parents, active participation in the provision of education to their children remains a formidable task. Many Traveller parents have had limited often negative experiences of the educational system themselves and consequently feel intimated when crossing the threshold of the school gates. As outlined in this book many Traveller parents estrangement from the educational process is compounded by a sense that they themselves are personally alienated from the school set-up and the perception that they are unskilled or unable to provide the support and supervision which their children need to succeed at school. A number of core themes emerge from this book, all of which impinge on the Traveller/school interface as it currently operates. They include:

Discrimination against Travellers
The Implementation of the Integration policy in the provision of Traveller education.
Withdrawal of Traveller children for learning support (especially during Irish language classes).
Promotion of Traveller culture by the education system.
Provision of relevant and ongoing education support for Traveller parents.
Relationship and communication between Traveller and school personnel.
Visibility of Traveller "voice" within the education system.
Inclusion of Traveller culture in school textbooks or syllabi.

The research sample for this project was drawn from five local Primary (first-level) schools in the Limerick city area, two of which are situated in areas designated as disadvantaged. The Traveller children interviewed for the research were living in a diverse range of accommodation including: in houses on a Traveller site; in chalets on a Traveller site; in trailers on site or in trailers on the side of the road. A quick review of the interview data as transcribed in this book makes it possible to highlight a number of culturally-specific aspects of Traveller life which still hold strong currency amongst the community. Most of the activities which Traveller children tended to participate in outside of school tended to relate to the Traveller community, extended family and the maintenance of Traveller culture. The training and management of horses and family and childcare tasks were much higher on the day-to-day agenda for Traveller children as compared with children of a similar age from the "settled" community. For many Traveller children their sense of self appeared to be strongly related to their ability to perform certain tasks within the extended family.

These tasks normally involved some degree of responsibility and were more often than not central to Traveller life and culture, a fact which was remarked upon by some of the Traveller researchers themselves as they undertook the project. The researchers discovered that for many Traveller children their overall sense of self-worth was, as with many non-Traveller children, strongly dependant on their parents' attitudes towards them. This interaction was slightly different from the same process as it occurs in the settled community as in many cases, Traveller children were more likely to elicit praise from their parents for *what* they could do (or indeed not do) as opposed to who they *were.* A strong respect for the family was very evident in the responses of all the Traveller children who engaged with the project and family in Traveller culture was clear in encompassing extended family and in particular, grandparents. Many children pointed to the strong affinity they had with their grandparents and the large role played by grandparents and the elderly generally within the Traveller community. As one researcher pointed out – "the children would be murdered for giving back cheek to a grandparent". This would certainly contrast with the inter-generational experience of many "settled" children where the placing of the elderly in nursing home care has become increasingly common. Other aspects of Traveller culture as recounted by Traveller children - some of which would differ as to their extent and significance amongst a peer group in the settled community – also revolved around the elderly to one extent or another. For instance, attendance at Anniversary Masses, prayers for the dead, for the sick and the visiting of holy places or holy people were all highlighted by Traveller children as an integral part of their social and religious formation. Children in most Traveller families were expected to participate fully in religious family events, many of which also involved the elderly members of their community. The majority of children expressed negative views about where they lived, a fact which is indicative of the huge ongoing challenges that still remain as regards the provision of adequate and appropriate accommodation for Travellers in Limerick and most other Irish cities. Unsurprisingly, many Travellers in Limerick have sought accommodation in areas that facilitate involvement in "traditional" Traveller activities such as horse-trading and scrap. Few local authorities in Ireland however provide facilities which include accommodation for scrap storage, horse management or living quarters for the extended family. As a consequence, the Traveller childrens' responses as relating to accommodation highlighted the "dirt", overcrowding and danger that they associated with the environments around the trailers and chalets where they lived. Preference for a particular type of home varied among the children. Some children indicated a preference for living in a trailer particularly as this mode of living made it easier to move and travel when necessary. These children were in the minority however and the experience of travelling was unfamiliar to many. This was not surprising in view of the absence of transient sites for Travellers in Ireland and the "boulder" policy that prevents most traditional Traveller stopping places from being accessed. Interestingly, when describing his identity, one boy said he considered himself a Traveller but described his family as "settled" for the simple reason that "we don't travel round anymore". Some of the interviews led Traveller children to discuss what they would like to be when they grew up and it was apparent from the responses that most children wished to follow in the footsteps of their parents and grandparents. For example, one boy stated that he wanted to have "loads of horses", so many that he would be a "farmer of horses", the same boy being adamant that he did not wish to own lots of land and thereby be a "farmer" in the traditional meaning of the word. Aspiring to motherhood remained very high on the list of life's milestones as outlined by Traveller girls indicating a contentment with the role model set by Traveller mothers within the community. The traditional desire of a young girl to become well-skilled in caring for the family, find a good husband and have a family of her own was reflected in the responses of many Traveller girls, some of whom already had considerable and responsible roles within the community as regards the care of the elderly including grandparents. The welfare of siblings, parents and grandparents was still seen as a essential role as fulfilled by the daughters and daughter-in-laws within Traveller families. In some of the interviews, conversations around the notion of friendships in the school environment developed into a discussion of relationships between the sexes within the Travelling community. The responses from the Traveller children indicated a strong awareness of what was deemed appropriate behaviour and emphasized a strong moral code which would have been more prevalent amongst their settled peers in times past. The children who brought up this topic described the importance of maintaining an appropriate distance between Traveller boys and girls, particularly as children approached their early teens. Completely "free" interaction between Traveller boys and girls was deemed unacceptable by most respondents and many children pointed out the preference for Traveller children from one site to interact only with the boys and girls from their own sites as opposed to other Traveller sites. Play between the sexes was discouraged unless the children were related to one another and Traveller girls were told by their mothers to – (in the words of the Traveller researchers) – "stay away from the boys" and to mind their "character". Mixing with non-family boys would incur disapproval from many Traveller parents and as one child stated, a girl would get "a bad name". As a Traveller girl's future could depend on her "reputation", a daughter would normally be chastised by her father or mother if she was playing with "bauld boys". The experience of Traveller boys was very different from that of the girls with boys generally being encouraged to branch out into the "men's world" of scrap and horses from an early age where they would associate "only with the boys" from then on. Some boys did mention however that it was in this male world of the Traveller men that they first learned how to "spot the girls"! The researchers noted that the children's responses echoed traditional conceptions of the gender/employment roles witin the Traveller community, this despite the fact that changing gender roles and high male unemployment levels within the general community have already impacted to some degree on traditional Traveller values and their way of life. The main findings of this research project, as elucidated by the Traveller researchers make for interesting reading and would be worthy of consideration by any future policy-makers in Ireland's statutory bodies, local authorities and amongst Traveller organizations at all level. Of particular concern and as highlighted by the Traveller researchers on a number of occasions is the failure of the Irish primary school system to meet their children's needs and the disengagement from the educational process that this has engendered amongst many Traveller families. The primary thematics findings of this research project can be summarised as follows:

That family was central to the Traveller children's life.
That the Traveller children took on responsibilities in the home.
That school was not talked about enthusiastically by the Traveller children.
That the school day was perceived by the Traveller children as long and boring
That teachers were most liked when they were fair, let the children out to play and did activity related subjects with them.
That some Traveller boys' relationships with the school principal, in the main, revolved around reprimand and punishment.
That both the Traveller boys and the Traveller girls talked passionately about horses.
That the Traveller children were hesitant to explore topics of a personal nature.

That the school world did not appear to be connected with the Traveller child's world.
That the Traveller children wished for improved standards of living accommodation.

Conclusion

The authors of this book acknowledge that the fact that the qualitative nature of their project meant that the sample of parents and children who worked on the project which culminated in *Whiddin to the Gauras* on a continuous basis was a small one. There is no doubt however that the experiences outlined in this book and the conclusions that can be drawn from these experiences will resonate with Travellers and other minorities on a much greater scale and well beyond the educational precincts of Limerick city. Perhaps the most fundamental conclusion as outlined in this book and the most valuable is its elucidation of what happens to a community of people who have to hide their true identity in order to engage with a so-called tolerant and multicultural society as modern Ireland would proclaim itself to be. The Irish government's **1995 Task Force on Travellers** was tasked with the job of charting policy and future objectives as pertaining to the Irish Traveller community. One objective which all Traveller and Roma and Sinti groups across Europe would endorse as an initial step forward would be the simple acknowledgement of Traveller and Gypsy identity. As this book makes abundantly clear there is little hope for progress in any societal sphere including education until the identity denial, which has pervaded many aspects of Irish life as applied to the Traveller community is negated once and for all. Identity denial as a discriminatory tool has no place in good educational practice anywhere and can only compound already existent problems that traverse a wide range of domains as relating to social disadvantage, drug abuse, lack of self-esteem and poor social support. No amount of books and reports can replicate the violence that is inflicted on a people who are labeled a problem from the moment of their birth. The Traveller children interviewed in this book have given an eloquent account of what it is like to watch your parents deny their own identity in order to apply for a job or go into a pub. They have hinted at the feelings of self-inferiority that can overcome a young person who is aware that their entire community is considered dirty and troublesome or who have to remove their earrings before attending a school in order that they are not singled out for discriminatory behaviour. That Traveller children are aware of the "distinction" that is made between their own community and the "majority" community at an extremely early age is one of the most poignant messages to emerge from *Whiddin to the Gauras*.

Acknowledgements

A special word of thanks is extended to both Eleanor Gormally and Mags Casey for their input and help when preparing the above essay.

Whiddin to the Gauras—Talking to our Own – (Traveller Researchers Talk to Limerick Traveller Children) (2005) (Authors: Mags Casey, Bridgie O'Donoghue, Ann O'Donoghue, Ann O'Driscoll, Eleanor Gormally) is published by Veritas Publications, Dublin – www.veritas.ie Email: publications@veritas.ie

References

Cauley, W. and Ó hAodha, M. (eds.) (2004). *The Candlelight Painter - The life and work of Willy Cauley, Traveller, Painter and Poet*; Dublin: A. and A. Farmar Publishing

Collins, M. (1994) *"The Sub-Culture of Poverty - A Response to McCarthy."* in *Irish Travellers: Culture and Ethnicity*. S. Ó. Síocháin et al. (eds.) Belfast: The Institute of Irish Studies, The Queens University of Belfast

Cottaar, A. et al. (eds.) (1998) *Gypsies and Other Itinerant Groups - A Socio-Historical Approach*; London: Macmillan Press

Court, A. (1985) *Puck of the Droms: the lives and literature of the Irish tinkers*. California: University of California Press

Dunne, P. and Ó hAodha, M. (eds.) (2004) *Parley-Poet and Chanter - A Life of Pecker Dunne*; Dublin: A. and A. Farmar Publishing

Harper, J. (1977) *The Irish Travellers of Georgia* M.A. University of Georgia

Helleiner, J. (2000) *Irish Travellers : Racism and the Politics of Culture*; Toronto: University of Toronto Press

Hyland, J. (ed.) (1993) *Do you Know Us at All?;* Dublin: Parish of the Travelling People

Joyce, N. (1985) *Traveller: an autobiography*; Dublin: Gill and Macmillan

Kenny, M. (1996). *The Routes of Resistance - Travellers and Second-level Schooling*. PhD. Trinity College Dublin

Kenrick, D. and Puxon, G. (eds.) (1972) *The Destiny of Europe's Gypsies*; London: Sussex University Press in association with Heinemann

Lee, J. (1989) *Politics and Society in Ireland 1912-1985*; Cambridge: Cambridge University Press

McCarthy, P. (1994) *"The Sub-Culture of Poverty Reconsidered."* in S. Ò. Síocháin et al. (eds.) *Irish Travellers: Culture and Ethnicity*; Belfast: Institute of Irish Studies, Queen's University

McDonagh, M. (2000) *"Nomadism."* in E. Sheehan (ed.) *Travellers - Citizens of Ireland*; Dublin: Parish of the Travelling People

CHAPTER FIVE

"DOING IT FOR THEMSELVES": IRISH TRAVELLERS AND THEIR LITERARY VOICE

MÍCHEÁL Ó HAODHA

In Ireland, as with other European countries, Travelling people have frequently been an object of fascination on the part of writers and artists. One need only take a quick glance at Ireland's literary output during the nineteenth century - the Irish Literary Revivalists including W.B. Yeats, Gregory, Synge or the writings of twentieth-century writers such as Bryan MacMahon, Pádraic Ó Conaire, James Stephens, Liam O'Flaherty, John B. Keane and Richard Murphy. It has generally been the case however that the image of Travellers as portrayed Irish literature, and as reproduced by primarily non-Traveller writers, has been made to fit a generic image. This image is in many ways the mirror-version of the colonial "stage Irishman", a character whose type is "constructed" of traits incorporating one or all of the following: the petty criminal, the happy-go-lucky vagrant, the drunk, the storyteller and the outcast.

This often-exoticist or "romantic" tradition is a relative exception however when compared with the manner in which Travellers – until very recently at least - tended to be consistently vilified and "Othered" in the public consciousness that is the Irish media. Indeed it arguable that much of the representations and imagery concerning Irish Travellers as exemplified in the modern-day media frequently differ little from the inferiorized or degraded discourse that was prevalent in aspects of the folklore tradition of the past. This essay discusses the emergence of that new genre that is Traveller literature and its impacts in terms of both advocacy and the articulation of Travellers unique community and cultural heritage. It also outlines the possible impacts of such literature to counteract or negate long-held negative perceptions and stereotypes which have had generations of reductionist stereotypes to draw on at this stage. The past decade has witnessed a frequent trumpeting of a "new" and multicultural Ireland, particularly on the part of the organs of the Irish state and the "liberal" intelligentsia, an Ireland where mutual tolerance and intercultural dialogue is the norm, a tolerance which bolsters the country's recent economic advances, advances which suddenly usurped decades of poverty, unemployment and mass emigration. An analysis of the manner in which particular minorities are "constructed" or represented shows that the "new" and vibrant Ireland may not be so different from the old. Breathnach (1998), who studied a representative example of Irish media reports as they referred to the Irish Traveller community between the 1960s and 1990s found a substantial increase in the number of reports that discussed the Traveller community in terms of race, discrimination and dialogue. While the new acknowledgement of a "sense of difference" as applied to the Traveller community was to be lauded she found a strong emphasis on sensationalist reporting which gave undue prominence to Travellers in terms of crime and violence. Increased representation of Travellers and acknowledgement of cultural difference was undermined by the continuance of Traveller stereotyping and a decontextualisation of the "real" issues common to tabloid journalism, that only served to reinforce prejudice and increase hostility towards Travellers as a community. She found that the issues of crime and violence as applied to the Traveller community continued to draw on a well of reductionist imagery that has been drawn upon for centuries. This reductionism was linked to a new fracturing of Irish identity along ethnic lines, one where "difference" as expressed between communities was increasingly viewed as undesirable and irreconcilable. This perceived sense of "difference" between the Traveller and "settled" communities was exacerbated by a lack of personal contact between both communities, as compared with the situation in a previously "rural" Ireland. The government task force entitled the *Travelling People's Review Body* (1983) and the *Citizen Traveller* campaign (2001) reinforced this perception when they both pointed to increased isolation of the Traveller community as a consequence of rapid social change and the "unfounded fears" generated within the settled community by this same lack of interaction between both communities. It is no small irony that a similar sense of "difference" based on a perceived lack of cultural compatibility is currently being used in relation to many immigrant groups by those elements in Irish society who would deny full citizenship rights and privileges to Ireland's increasing immigrant population. How has the increasing body of Traveller-produced literature and autobiography reacted to this changing situation encompassing new threats and types of discrimination which are frequently more subtle or difficult "to pin down"?

While media and "settled" community perceptions of Travellers have a somewhat longer history the public articulation of Traveller identity on the part of the Irish Traveller community itself, is a relatively recent phenomenon. It is a phenomenon that has focussed in particular on a number of core aspects of the Traveller experience. These aspects are ones which Travellers themselves have identified as contributing to the deliberate erosion of their culture when operating in tandem with an overtly assimilationist ethos as directed towards their community. One extremely important element in this new articulation of identity on the part of Travellers themselves is the increase in Traveller self-representation through the mode of autobiography. Travellers' liminal position in Irish society has commonly meant that representations of them had largely been effected from without. However recent decades have seen the emergence of an increasing number of powerful evocations of Traveller identity from within the community itself. This canon of work is one where the voices of Travellers themselves are heard for the first time in the public sphere and where their culture is celebrated in a unique and intimate fashion. These evocations are all the more remarkable when one considers the high levels of non-literacy, social exclusion and the denial of access to the dominant means of self-representation that has been characteristic of Traveller life until very recently. A major aspect of the more recent Traveller literature which has emerged is the manner in which it challenges the static and stereotypical way in which Travellers have been reproduced culturally, politically and ideologically in the past. A strong underlying tenet of this "new" literature is the attempt to forge a "place" for Travellers in modern Ireland, a place that is no longer dismissed or assigned to the margins. It can be argued that Irish Travellers have in the past had little consciousness of their culture and language as distinctive attributes. These publications are an example of a fundamental change in this regard however. A number of common themes have emerged in these texts that form the core of this challenge and which seek to undercut the common view of Travellers' perceived liminal position in Irish culture. The new voices that are to heard in these publications undoubtedly form an important challenge to those "official" or written histories of Ireland which have ignored groups whose culture has been primarily oral. They also function to forcefully reject the "drop-out" theory of origin that continues to haunt much modern-day representations and treatment of Travellers. Other themes which emerge strongly in this autobiographical work by Travellers include a focus on education and inclusion in Irish society on Travellers' own terms, a tendency which is also accompanied by an uneasy presience and fear of cultural annihilation. Evident too is the lingering anxiety that many reductionist stereotypes as applied to Travellers in the past have been internalized by the Traveller community itself, in particular younger Travelling people, thereby damaging irreparably their cultural self-identity. A strong emphasis is also placed on documenting the instances of abuse and discrimination which have led to Travellers exclusion from Irish society.

These writers frequently express a bewilderment at the levels of prejudice and hostility exhibited by the Irish "settled" community towards Travellers and exhibit a longing for the acceptance of a nomadism, now frequently referred to as the "nomadic mindset" which is the kernel of Traveller identity. Books such as Seán Maher's *The Road to God Knows Where* (1972), Nan Joyce's *Traveller* (1985), Pecker Dunne's *Parley-Poet and Chanter* (2004) and William Cauley's *The Candlelight Painter* (2005) are all very aware of their quasi-oral status, one which generates a unique and creative tension between print and speech. Both Maher's and Dunne's books are littered with folk-anecdotes, songs and superstitions all of which highlight the educational function that the oral tradition occupies within the Traveller community as regards the cultural transmissson of knowledge. They indicate the high value placed by the Traveller community on the skills of the storyteller who can effectively disseminate Traveller values and historical tradition. The oral tradition as elucidated in these narratives functions as a coping mechanism whereby the past can be communicated with and the present can be dealt with as Maher (1972) makes clear:

> …you must remember that storytelling is our only means of communication with the past. We on the road can't write our deeds, but we do remember them and pass them down faithfully by word of mouth (1972, 68).

These books expose a range of common stereotypes about Travellers that continue to have a strong resonance in Irish public opinion. One example of this type of rebuttal, for instance, is the under-cutting of the long-standing stereotype that Travellers were always nomadic and never lived a sedentary existence for any significant length of time. The usurpation of a Traveller stereotype such as this is but one example the importance of which cannot be underestimated when one considers that much assmilationist policy as directed towards Travellers has assumed a permanent form of nomadism to be the Traveller norm. Maher, Dunne and Cauley all provide rebuttals to this common misconception about Traveller nomadism with Maher (1972) describing his family's nomadism as seasonal in nature:

> Most of the bad winters, my parents would abandon the tent or the wagon and move into a spike for a few months. In those days there was one in every large town in Ireland…I liked the spikes very much during the cold winter nights, for I could sit at the turf fire with my father and all the old men…Night after night in the spike he would hold me on his lap near the blazing fire and sing and tell stories to the other men there (1972 9-10).

An emphasis on the importance of education and literacy to Travellers is also a theme which emerges in this autobiographical canon with both Maher and Joyce expounding on their struggles in this regard and the importance of this aspect of human dignity to the articulation of a community's "true voice":

> It's an awful sad thing to have no education…If I knew how to write properly and had good spelling I wouldn't have done this book on tape. Anna Farmar [the book's editor] recorded it and wrote it out for me – if I wrote a book myself whoever published it would need a medal for bravery! (Joyce 1985, 95)

The mediation between print and speech which serves to emphasise the orality central to Traveller culture is given a new awareness by the inclusion of various Shelta words in these autobiographical works, a pattern that has been given added impetus in some of the most recently-published Traveller literature including Peter Brady's novel *Paveewhack* (2001) and, latterly, in the works of Cauley (2004) and Dunne (2004) . It can be argued that the written use of Shelta, a language into which very little research has been undertaken and which appears in the past at least to have had a strong secrecy function is "political" in the sense that it serves to highlight the "sense of difference" that is felt by a community who, when it has been convenient, have often been dismissed as having no fundamental cultural attributes which distinguish them from the "settled community" in Ireland. A significant increase in the use of Shelta in the most recent Traveller publications may also be indicative of the huge assimilative pressure that the small minority that is the Traveller community is currently experiencing alongside the frantic pace of societal change that is affecting all members of the Irish population, be they Traveller or non-Traveller. It may also be indicative of the feeling of foreboding that pervades the texts of these Traveller writers, who, it should be noted, are elderly in terms of their own community, a foreboding foregrounded in the fear of their possible cultural disappearance as a separate group in Ireland. Maher, Cauley and Dunne all make reference to the weakening role of the Traveller language (languages) cultural marker within the Traveller community, a fact which cannot be separated from the decline of many "minority" languages in an increasingly globalised world, but one which in the context of the Travelling community's history cannot be divorced from the sustained undermining of Traveller identity that has accompanied the modernisation of Ireland. Maher, whose autobiography was published over forty years ago, includes a brief glossary of Shelta words at the end of his book, a glossary which is prefaced with the warning that this language is "now almost-extinct" while Dunne and Cauley preface the different sections of their books with phrases in their respective Cants complete with English explanations, including for example:

> The drom in the clammy pani – *On the road in bad weather* – (Dunne 2004, 19)

> Our dhíls scurlimed - *Burning us out* - (Cauley 2004, 26)

Dunne (2004) warns of an absence of interest in this aspect of Traveller culture amongst the younger generation and links this with the damage that has been inflicted on Traveller self-identity and the transference of Irish identity as a whole:

> The way I see it the Travellers here don't speak the Cant very much anymore. More's the pity. That's what I say. They are letting it die you see. A lot of Travellers in Ireland are not speaking our own Cant because they are ashamed of it. They don't want to be called the "knacker". I don't understand this. They should be proud of who they are and who their ancestors were. It's a bit like those people in the West of Ireland who have lovely Irish but who won't use it or pass it onto their children for fear that people might think they are ignorant. It's a lovely thing to know the Cant and it is a very old language (2004, 51).

While these autobiographies are extremely positive in many respects and include many vignettes which emphasise the joy, resourcefulness and adaptability of Travellers in often-adverse circumstances, the books also intimate that the threat of language death may be indicative of a greater cultural disappearance. The ambivalence in the title of Maher's book *The Road to God Knows Where* (1972) is followed by apocalyptic "warnings" that the road may not lead anywhere or that the destination is ultimately one of assimilation or cultural disappearance.

> You see, son, life is becoming more difficult and the road as a way of life is finished…(1972, 100).

> The time shall come too, brothers and sisters when the travelling people of Ireland shall be no more (1972, 78).

It is the death-knell of the most defining Traveller cultural attribute, nomadism, that which is inherent in the very nomenclature by which Travellers refer to themselves – Travellers – that many of the writers identify as the most serious threat facing their community. Both Maher (1972) and Joyce (1985) identify the disjuncture between the Traveller and "settled" views of nomadism as the roots of much of the prejudice that has been indicative of Settled-Traveller attitudes and relations:

> We lived in a house for a very short period when I was young, but my parents could not adapt to settled life. In the town where we lived we were not wanted; there was always a great amount of prejudice among the townspeople, especially those who lived near us. For my father it was not too bad because he went out the country each day with the pony and cart. For my mother, however, things were different. She was always alone in the house; none of the neighbours would even talk to her – let alone come into our house for a visit or a cup of tea. To the townspeople we were dirty begging tinkers and no respectable person would visit us…(Maher 1972, 15).

> My third little boy, Patrick, was born in St. James's and he was christened when we moved over to Finglas. We stayed there for about two years until Elizabeth was born but the Corporation kept shifting us, from one field to another. People say to us, "Will you not settle down?" but we don't get the chance because we're hunted from place to place… (Joyce 1985, 80)

Joyce also describes the activities of the police and vigilante groups in Northern Ireland where Travellers were doubly prejudiced-against for sectarian reasons:

> Some of the police were very hard on us. They would come up at ten o'clock of a winter's night and it didn't matter if it was snowing or freezing, the travelers would have to pack up. The women might have walked miles for two buckets of water to cook the food and wash the children. They might have spent an hour trying to light the fire…If you had any brain at all you knew that the travelers weren't wanted. Once in the North we stayed with two families for about a week, near Warrenpoint. One man's horse jumped through a gap into a farmer's field. A group of men came up about twelve o'clock of a winter's night with shotguns that they use for hunting foxes and rabbits. They shot this man in the legs and he was in terrible pain. We shifted but it was so dark we didn't know where we were going. (1985, 40).

While Joyce highlighted the phenomenon of eviction, an aspect of Traveller life which has recurred again with increased regularity since the introduction of the *Housing Miscellaneous Provisions Act* (1992), Cauley (2004), a landscape painter by profession, has written of the psychological difficulties induced by Travellers' forced settlement in Irleand in the absence of transient sites or any provision for those who wish to remain nomadic and the destruction of Travellers' "collective memory" through the deliberate blocking off of traditional campsites that had been used regularly by generations of the same family:

> I wonder how many older Travellers became depressed because they weren't allowed to travel anymore? How many of them became depressed when they saw their traditional camping places being blocked up and they ended up drifting from the countryside into houses and the cities?... A number of my siblings who are now dead suffered from depression while they were alive. I think the city life may have had a lot to do with it. Freedom and travelling was in their blood and being cooped up in the city affected them and made them sick. I have heard that the same thing has happened to many other traditionally nomadic people who have been forced by circumstance to live in the cities – people like the Native American Indians and the Aborigines in Australia... Someday I will try and capture this misery in a painting. I will get a huge canvas, the biggest one I can find, and when the painting is done I will show it to the press and the government so as to make a statement to them. "Look it, this is how the government ignored us back through time. This is how you hurt these peoples' minds then, a hurt that later changed into a depression. At a later stage you forced them into houses. You blocked up their roads. I will put all my energies into that painting. Everything *(2004 23-41).*

In the same book Cauley compares his identity as grounded in a nomadic tradition with a nature that has been befouled by rapid and often-forced social change:

> I was often asked if I was born again what would I like to be. Well, the first thing that comes to my mind is a swan. Yes, I know it's a strange thing to say but I have my own reasons for that answer. When you look at how peaceful a creature the swan is it seems to have no worries, it comes and goes as it pleases. And most of all it has rights. No one tells it to move on, no one tries to control it *(2004, 109).*

The importance of nomadism (now increasingly referred to in the literature as "the nomadic mindset") even to younger Travellers who have never had the same opportunities to travel as their parents and grandparents is powerfully highlighted by a young Traveller named Martina Joyce interviewed for a series of essays on Traveller life and culture and produced as Traveller Ways, Traveller Words (1992) by Pavee Point, a Dublin-based voluntary organization and partnership comprising both Travellers and non-Travellers committed to the attainment of human rights for Irish Travellers. She expresses a longing for a nomadism and highlights the alienation and lack of social contact that accompanies enforced settlement:

> You get ta see more an' ya meet up with different Travellers that you never seen before. You mix better. You wouldn't be always around your own that ya do be around in the sites and that. You're always nearly with your own people. Mixin' with the extended families and that. And ya could pick up things and do things like. Pick up like doin' things in markets an' that. Travel to the markets an' the fairs an' things like that. Out sellin' in houses an' things. Well, nowadays you don't have the freedom to travel. Like, if you go to the camps most o' them are all bowldered up with stones in front o' them an' that. The sides o' the roads is all done. And then if you go tiv a site nearly all the sites is full up because there again the wans that's in it can't go movin'. And then if y'are aan the side iv a road or in a field you're put out o' them mostly with the guards an' the people who owns the land. I don't know, if you've no skips or an'hin' to put the rubbish in they'll say that you're pilin' the rubbish an' you're gatherin' dirt an' this, that an' the other an' you're ruinin' the countryside an' they have ta put ya goin. An' then when you go, then, you've nowhere else ta go because all the sites are full up with the Travellers that's 'ithin in them. So wha' else will ya do? You'll only have ta go because they'll drag you out of it if ya don't go. (1992, 72-73).

Much Traveller literature expresses an incomprehension at the level of prejudice that is often evinced by the "settled" community towards Travellers, a prejudice, the roots of which frequently lie in ingnorance and fear. Both Dunne and Cauley describe an anti-Traveller prejudice that blighted their memories of childhood:

> It is funny but no matter how hard you try you will remember to your dying day those moments when people were prejudiced towards you. Every human being wants to be respected…The worst part of prejudice like this is when your childhood innocence is destroyed by it. I remember seeing people shouting at us when I was a child. "The Tinker isn't wanted here, go away", and all the rest of it. They made a distinction between us and the rest of the Irish people. That's the best way to describe it. And you were aware of that distinction from the moment you were born. I knew what it was from the moment I was able to walk. The badness in some people…This prejudice is an old thing in Ireland and there isn't any sign that it's going to go away. I was talking to a few fellas there recently. They are Gypsies, travelling people from from Eastern Europe. Roma or

> Romanies is what they call themselves. I was asking them what they thought of Ireland and one of first things they said to me was "My God, but this is an awful racist country". "Tell me more about it" says I to them. "If anybody knows about the racism here, it is me". "I have seen it since I was born, since I came from my mother. Against me and people like me. And I can't explain fully what the reason is behind it".

Maher (1972) admits that his interest in literacy and the thirst for a "formal" education which tormented him for much of his youth, was as much a means of attempting to understand the prejudice against Travellers as it is a natural inclination for learning:

> You know", I said, I keep thinking about people. I keep asking myself, why do people have to hate each other? I keep thinking how nice it would be and how happy everyone would be if they did not hate each other. I don't know why I keep thinking like this; perhaps my father is right when he says that I'm a funny one. There are times when I go and beg from house to house for old books (1972, 101).

Cauley (2004) expresses his bitterness at the intolerable burden that is the prejudice against his community by taking refuge in his imagination. He describes one of his occasional "flights of fancy" which functions as a form of escapism from a reality that is a good deal more grim:

> Sometimes I feel very bitter about the way Travellers were treated in Ireland When I was younger I sometimes hoped that I could just disappear out of this country. In my mind I used to wish that I someone would kidnap me so that I could escape from some of this hardship!..(2004, 74-93).

Conclusion

Taken as a whole, this entire Traveller autobiographical canon can be seen as a challenge to the manner in which Travellers have traditionally been precluded from discussions of modern Ireland. This is unsurprising given that many canonical historical texts, by both nationalist and revisionist historians alike, have simply excluded all mention of Ireland's oldest minority. In the very few cases that Travellers are alluded to at all they tend to be depicted primarily in the role of passive recipients of the historical process – a group whose perceived position as drop-outs – i.e. victims of colonial eviction and/or the Great Famine rendered them uninterested in such issues as Irish nationalism, the fight for Irish Independence or uninvolved in Irish history generally. As a consequence perhaps, Traveller authors have been at pains to emphasize their Irishness and their critical role in the development of Irish cultural processes generally.

In conclusion it is worth returning to the work of Nan Joyce (1985), the first Irish Traveller ever to fight on behalf of her community on a national and political level in Ireland when she ran as a political candidate on a Travellers rights ticket in the the Dublin of the early 1980s. She situates her autobiography within a context whereby a good deal of Irish anti-Traveller prejudice can be firmly identified as having its roots in ignorance. Her description of a public meeting - in 1985 - with a group of Irish University students is an unbelievable indictment of the levels of ignorance that exist regarding Traveller culture in an Ireland where stereotypes have become so reified as to almost assume the status of "truth":

> Some of the students – and they were Irish – knew nothing about the travellers, it was as if we'd if we'd come from the moon. The questions they asked! "Do you marry?" They didn't know if we got married or if we lived together, if we had the children on the roadside of if we went to hospital. It was like speaking to something from out of the world. They could tell you the history of the North American Indians but here were the Irish travellers in their own country and they could tell you nothing about them, their history or their traditions. And these are the people who'll be running the country when we're all gone. (1985, 108)

She calls both for younger Travellers to jettison their feelings of inferiority and for the "settled" Irish community to make themselves more aware of the minority in their midst:

> I would like all the travelling children to have self-confidence and to grow up proud of what they are because they are very special people with their own traditions and their own way of life. But the way they've been treated and discriminated against they grow up ashamed of their own parents. It used to be that the one time you heard about travellers was when they did wrong; nothing about us was ever taught in schools and when children came by our camp they would just run by – they were afeared of the travellers. Even today some people are afeared to come near us and chat with us – it's a fear of the unknown. (1985, 118).

References

Acton, T. (1974) *Gypsy Politics and Social Change*; London: Routledge and Kegan Paul

—. (1994) "Categorising Irish Travellers" *Irish Travellers: Culture and*

Ethnicity in S. Ó Síocháin et al. (eds.) Belfast: Institute of Irish Studies, Queen's University of Belfast

Cauley, W. and Ó hAodha, M. (eds.) (2004). *The Candlelight Painter - The life and work of Willy Cauley, Traveller, Painter and Poet*; Dublin: A. and A. Farmar

Collins, M. (1994) *"The Sub-Culture of Poverty - A Response to McCarthy."* in *Irish Travellers: Culture and Ethnicity*. S. Ó. Síocháin et al. (eds.) Belfast: The Institute of Irish Studies, The Queens University of Belfast

Donahue, N. and Gmelch, S. (eds.) (1986) *Nan: the Life of an Irish Travelling Woman;* London : Souvenir

Dunne, P. and Ó hAodha, M. (eds.) (2004) *Parley-Poet and Chanter - A Life of Pecker Dunne*; Dublin: A. and A. Farmar Publishing

Geertz, C. (1986) *The Uses of Diversity*; Massachusetts: Tanner

Helleiner, J. (2000*) Irish Travellers : Racism and the Politics of Culture*; Toronto: University of Toronto Press

Kenny, M. (1996). *The Routes of Resistance - Travellers and Second-level Schooling.* PhD. Trinity College Dublin

Lentin, Ronit and Robbie McVeigh (eds.) 2002. *Racism and Antiracism in Ireland.* Belfast: Beyond the Pale Publications

Liégeois, J-P. (1994) *Roma, Gypsies, Travellers;* Strasbourg: Council of Europe

Mac Gréine, P. (1931) *"Irish Tinkers or "Travellers", Some Notes on their Manners and Customs, and their Secret Language or "Cant". Béaloideas: the Journal of the Folklore of Ireland Society* Vol. 3: 170-181

—. (1932) *"Irish Tinkers or "Travellers"." Béaloideas: the Journal of the Folklore of Ireland Society* Vol. 3(2): 170-186

—. (1932). *"Further Notes on Tinkers' "Cant"." Béaloideas: the Journal of the Folklore of Ireland Society* 3(3): 290-303

—. (1934). *"Some Notes on Tinkers and their "Cant"." Béaloideas: the Journal of the Folklore of Ireland Society* 4(3): 259-63

Mac Laughlin, J. (1995) *Travellers and Ireland: whose country, whose history?;* Cork: Cork University Press

—. (1996) *Travellers are still victims of a Victorian image of society Sunday Tribune – 12/9/1986*

McVeigh, R. (1992a) *"The Specificity of Irish Racism" Race and Class.* Vol. 33, No. 4. 31-45

—. (1992b) *Racism and Travelling People in Northern Ireland* – 17th Report of the Standing Advisory Commission on Human Rights

—. 1994 *Theorising sedentarism: the roots of anti-nomadism. [*Paper for

ESRC Romani Studies Seminar] University of Greenwich . Belfast: Campaign for Research and Documentation

—. (1997) *Theorising Sedentarism: the roots of anti-nomadism* in T. Acton (ed.) *Gypsy Politics and Traveller Identity*; UK: University of Hertfordshire Press Nicholson

O' Connell, J. (1994b) *"Ethnicity and Irish Travellers"* in S. Ó. Síocháin et al. (eds.) *Irish Travellers: Culture and Ethnicity*; Belfast: Institute of Irish Studies, Queen's University

—. (1992a) *"Working with Irish Travellers"* in *DTEG File* Dublin: Pavee Point Publications

—. (1994a) "*Reach Out: Report on the "Poverty 3" Programme", 1990-94.* Dublin: Pavee Point Publications

—. (1995) *"Travellers and History."* in N. Ní Laodhóg (ed.) *A Heritage Ahead. Cultural Action and Travellers;* Dublin: Pavee Point Publications

Okely, J. (1983) *The Traveller-Gypsies*; Cambridge: Cambridge University Press

Sibley, D. (1981) *Outsiders in urban society;* Oxford: Basil Blackwell

CHAPTER SIX

ROMA AND SELF-REPRESENTATION: SOME ASPECTS OF A ROMA ACTIVIST'S EXPERIENCE

ANA OPRIŞAN

In this essay, Ana Oprişan, who was born into a part-Roma/part-Romanian family in Romania, describes the process whereby she came to an understanding of her own particular part-Roma ethnic identity in adulthood. She also describes her work as a Roma activist in Turkey and gives some insights into the situation of various Roma communities in the Middle East, many of whom had relatively little contact with Roma organisations in Europe or within their own countries until very recently.

***

I was born in Romania to a Roma (Gypsy) father and a Romanian mother and lived in Romania until 2002. My father was brought up within the Roma community but left that "world" behind when he married his non-Roma wife – my mother. Like the members of many minorities worldwide my father became acculturated to the "majority" culture in Romania and tried to leave his original identity behind. He tried to forget about his family's identity and to "pass" within the majority culture. He believed that it was easier for himself and his wife and children to function and achieve socially within Romanian society if he denied where he had come from originally - in the process. Although this might seem to be an easy thing to do, in reality it is anything but easy to try and change your identity. Your identity or your background will always come to the surface at some point despite your best efforts. Ironically, although there is huge prejudice against the Roma in Romanian society, it wasn't in the "outside" world that the question of my father's and our family's part-Roma ethnicity first came to the surface. In fact, it was in the confines of the family home that we children first became aware of it. Both my sister and I were brought up as normal "advantaged" Romanians and our parents ensured that we were kept apart from all our relatives on my father's side of the family. Any discussion about where my father was from was a taboo subject in our house growing up.

I became more aware of my Roma heritage and the various "issues" associated with this when I first visited my grandparents' house in Frumuşiţa/Galaţi on the death of my grandfather (RIP). I was about seventeen years of age then and for the first time in my life I met a whole range of different people who were my blood-relatives. As with all local Romanian people I knew that the "ţigănia" was a place where only the Roma people lived. I knew that they were a people who were considered "different", "inferior" to everybody else but also dangerous, and that they had come to that place sometime in the past and settled there in a community that was very separate from all other Romanians. On the day of the funeral I discovered a little bit more about my grandparents through my conversations with the relatives at the funeral. My father's parents had been travelling throughout Romania and as with many nomadic people it wasn't clear whether they originally had a particular village or region which they had considered their "homeplace". My father had been born while they were "on the road" in a place called Bârlad / Vaslui which is in the North-East of Romania.

Roma Advocacy

As I became older I noticed that I had a certain facility of languages. I managed to find out about a programme that combined both my linguistic skills and aspects of the ethnic background I knew I somehow had and eventually became a student in the Romanian – Romani Languages Section of the Arts Faculty in the University of Bucharest, in the capital of Romania. I graduated from this University in the summer of 2002 but had already started working on Roma-related issues about two years prior to my graduation. Between 1998 and 1999 I made contact with the *Roma Centre for Public Policies* – also known as *Aven Amentza* – in Bucharest where I began to work on a number of Roma-related policy-making projects and advocacy issues. It was an exciting time to work here and to witness at first hand the evolution of the Roma "Movement" and their various advocacy initiatives in Romania. Until 2002, my part-Roma ethnicity was not really a major "issue" for me personally or for those others who were working with me on these various Roma-related projects. The debates that sometimes take place between Roma activists and non-Roma or part-Roma activists whose work concerns Roma-related issues and rights was not an "issue" for me as relating to my own background.

As with Roma and Traveller-rights organisations in many other countries the "Who is Roma and who is not Roma?" question or "dilemma" is still a "contested" issue in the Romania Roma movement, one which can sometimes cause "tensions" as relating to representation and advocacy work on behalf of different cultural groups.

For me all this experience was also like an effort to understand the Roma reality and myself in this new asserted context. Maybe unconsciously, the work I did or I was trying to do was a kind of a retribution / paying back for all that people like my father and my relatives who had to face difficulties because of what they were in a time when I lived unaware and carefree.

A defining moment for me, both on personal level and as relating to my career occurred when I first came into contact with the Turkish-speaking Muslim Roma in Romania - often referred to as the Horahane Roma. This Turkish-speaking community are, one can argue, a "minority within a minority". While they live in the same country as other Roma they nevertheless are considered "different" and are sometimes subject to prejudice both from non-Muslim Roma in Romania, other Turkish diaspora populations (groups often referred to as Turks and Tartars) and from the broader non-Roma society in Romania.

In fact my ensuing work with this community was one of the primary reasons why I decided to leave Romania "permanently" and move to Turkey or – "the place of the Muslim Roma". My interest in working with this group of Roma coincided with a small but burgeoning movements towards the social and cultural "awareness-raising" amongst the various communities of Roma in Turkey – including the Horahane Roma. A group of people which included myself came together who were interested in helping various the various Roma communities in Turkey agitate for their rights and who wished to create an organisation which could do so. This small group but enthusiastic group of people – a group of whom I am very proud to be a part of - started this advocacy work in Turkey with the launch of the *Romani Studies Network*, a network which combines both fieldwork, academic research with awareness-raising/advocacy work throughout a range of Turkish Roma communities as relating to social, cultural and educational issues.

Official estimates put the number of Roma in Turkey at 500,000 people. As with statistics relating to many Gypsy, nomadic and often-marginalised groups however, it is very difficult to know how accurate any of these estimates are. For instance, I have spoken to Roma leaders and spokespeople in Turkey who put the real figure for the Roma population in Turkey at 5 times higher than the estimated figure 2.5

million and there are others again who give different figures. Gypsy communities can be found throughout Turkey, and there are very significant Roma populations in the Thrace[1] and Marmara[2] regions (the Rom groups) and also in Anatolia[3] (Dom[4] groups) including areas such as Maras, Antep, Adana or Istanbul. Roma whose families originated in Bulgaria make their homes in areas such as Kayseri, Adana, Osmaniye, Sakarya and Çorum and as with Roma everywhere there are variants as regards group self-ascription and names/ascriptions as applied to these groups by non-Roma. In Çorum[5], in the north of Turkey, for example, the

---

[1] Today the name Thrace designates a region spread over southern Bulgaria (Northern Thrace), northeastern Greece (Western Thrace), and European Turkey (Eastern Thrace). Thrace borders on three seas: the Black Sea, the Aegean Sea and the Sea of Marmara. In Turkey, it is also called Rumeli. Ancient Thrace (i.e. the territory where ethnic Thracians lived) also included present day northern Bulgaria, north-eastern Greece and parts of eastern Serbia and eastern Republic of Macedonia. The European portion of Turkey, known as Trakya (Thrace), encompasses 3 % of the total area but is home to more than 10 % of the total population. Thrace is separated from the Asian portion of Turkey by the Istanbul Bogazi (Bosphorus Strait), the Marmara Denizi (Sea of Marmara), and the Çanakkale Bogazi (Dardanelles Strait).

[2] The *Sea of Marmara*, also known as the Sea of Marmora or the Marmara Sea, is an inland sea that connects the Black Sea to the Aegean Sea, thus separating the Asian part of Turkey from its European part. The Bosphorus strait connects it to the Black Sea and the Dardanelles strait to the Aegean. The former also separates Istanbul into its Asian side and European side.

[3] Anatolia is a large and, roughly-rectangular peninsula which is situated bridge-like between southeastern Europe and Asia. The Anatolian part of Turkey accounts for 97% of the country's area. It is also known as Asia Minor, Asiatic Turkey, the Anatolian Plateau. The term Anatolia is most frequently used in specific reference to the large, semiarid central plateau, which is rimmed by hills and mountains that in many places limit access to the fertile, densely settled coastal regions.

[4] The Dom (or Domi) of the Middle East are frequently considered a section of the larger Roma (or Gypsy) ethnic population. They may also be related to the Domba people of India. They have a rich oral tradition and express their culture and history through music, poetry and dance. Their estimated population of 2.2 million is primarily spread across parts of Turkey, Egypt, Greece, Jordan, Israel, Lebanon, Iraq and Iran. The actual population, however, is thought to be higher as some Dom are excluded from national census and others label themselves in national terms rather than as Dom. Nowadays, many Dom speak the dominant languages of their larger societies, but Domari, their national language, continues to be spoken by more insular communities.

[5] Çorum is a city with a population of nearly 180,000 people in the Çorum Province of northern Turkey. Historically, Çorum was associated with the famous Ertuğrul family. Their Karadeniz (Black Sea) dynasty dated back to the early 12th

"settled" Roma living there, whose families originated in Bulgaria are often referred to as *"Haymantos*", while in cities such as Erzurum, Artvin, Bayburt, Erzincan and Sivas the Roma are frequently given ascriptions such as *"Posha"* or *"Bosha"*. In Van, Hakkari, Mardin and Siirt, it is slightly different again and Roma in these areas are most often identified as *"Mutrib"* - the word *"mutrib"* or *"mırtıp"* meaning "musician" in Arabic.

Two basic observations can be made in the Turkish context as relating to both the hetero-identification of Roma / Gypsies (i.e. the identification of the Roma by non-Roma) and self-identification - as indicated by the Roma themselves. Roma in Turkey *refer to themselves* as "Roman" or "Çingene" – a self-identification as relating to their ethnicity - or by using only the religious self-identification, for instance – referring to themselves as *Muslim*, *Alevi* or *Abdal,* a form of reference which is quite common amongst those Roma who are Shia. The latter form of self-identification is based on religious ascription and avoids any reference to Roma origins. The names given to the Roma are wide-ranging and vary from city to city and region to region. They are sometimes *referred to* as *Posha* (in Eastern Anatolia especially) or as *Mırti/Mutrib* in places such as Hakkari, Mardin, Siirt and in the southern part of Van). In Adana and Osmaniye Roma are frequently referred to by the terms *Kocer and Cono* – (pronounced "djono") while in the capital of Turkey, Ankara, they are frequently called *Teber.* In Diyarbakır[6], south-eastern Turkey, the Roma[7] are known by a range of interesting names, some of which have not previously appeared in the literature relating to the populations and cultures of this area of Turkey. Some Roma here are referred to as *Kereçi / Karaci* (a Persian word meaning "stranger" or "foreigner") while others are referred to as *Ashik* or *Gelsin*. Christian Roma are referred to by a different name – i.e. *Balamoron* - although this term is limited to the identification of Greek

century and the family ruled in Çorum for seven centuries until the establishment of the Republic of Turkey in the early 20th century

[6] Diyarbakır is a major city in southeastern Turkey situated on the banks of the River Tigris, and is the seat of the Diyarbakır Province. It is the second-largest city in Turkey's Southeastern Anatolia region, after Gaziantep, with a population of over 500.000 people of which a large Kurdish population.

[7] Here, I use the term "Gypsy" and not "Roma" for the groups of South-Eastern Turkey and the Middle East in general, as they are not considered to be Roma. These groups are sometimes referred to as Dom or Lom or are known frequently given this ascription by others – both Gypsy and non-Gypsy. These latter groups often comprise other and varied nomadic groups whose mode of living is very similar to that of the stereotypical/archetypal "Roma" / "Gypsy" populations).

Christian Roma only. As with Roma in other countries a frequent method of Roma ascription/categorisation relates to the different trades which particular families and communities have traditionally followed – e.g. *Elekçi* (sieve-makers), *Sepetci* (baskets-makers), *Kalayci* (Tinsmith), *Demirci* (Ironsmith), *Kemikci* (Kokalara – artisans who work in bone), *Bohcaci* (door-to-door pedlars), *Arabaci* (horse-carriages operators) etc.

In a country as large as Turkey, it is no surprise to discover that religious allegiances are diverse, a fact which makes the identification of different Roma groups more complex, particularly given that religious affiliation has a very strong influence on the self-identification of many Roma communities.

Self-ascription and the identification of Roma by non-Roma. For instance, the *Posha* groups from the Van area of Eastern Turkey are considered to be Muslims, while those Roma living in Tokat[8], Eastern Turkey who are today frequently regarded as Muslims, identified themselves in the not-too-distant past as Christians. The Roma living in south-eastern Turkey appear to follow Islamic religious rites which are close to the religious beliefs of the Çuki, the Shia Muslims, the Alevi and the Abdal groups. It is also the case that some of the *Mırtip/Mırti/Mutrib* Roma, many of whom who live in Siirt, Hakkari, Mardin and Van are Muslims of the Shafi rite. It is generally acknowledged that the vast majority of the Roma in Turkey are Muslim and their religious affiliation lies amongst a range of "groups" including the *Shafi Sunni Shafi or Hanefi, the Shia, the Alevi and the Abdal)* [9]

Turkish Roma and the Future

Roma in Turkey face a similar set of social issues and problems in the present-day as many of their counterparts in both Western and Eastern Europe. Prominent concerns as expressed by Turkish Roma communities themselves include: employment, housing, education, health, legal provisions and prejudice in terms of society and the law. The fact that many large sections of the Turkish population live in very difficult social

[8] Tokat is a city in Anatolia. It lies inland of the central Black Sea region and is over 400 kilometres from Ankara, the Turkish capital.

[9] This statement is made with the proviso that little research has, as yet, been completed, as relating to the range and diversity of nomadic and "Gypsy-like" groups who live and work in South-eastern Anatolia and it often quite difficult to work out the exact religious affiliation of different groups.

and economic conditions today is often used as an excuse or pretext for not addressing the needs of minorities and ethnic groups within Turkey who are considered to inhabit the margins of society.

While the Roma in Turkey do not face persecution in the form of torture etc. due to their Roma ethnicity, it is undeniably the case that their position in society involves considerable mis-treatment, particularly as related to their categorization and perceived "Other" social status. The Roma, both historically, (and in the present-day) - are most often mentioned in terms which are negative and reductionist. They almost inevitably appear in regional and national ordinances and laws which mention such issues as – e.g. "dealing with delinquency", "dealing with beggars", "preventing illegal acts" etc. as elucidated in official and state sources. While Turkish Roma have not faced the ugliness of ethnic cleansing and racially-motivated attacks as Roma have in many European countries, one only needs to scratch under the surface of the "romantic/folkloric" toleration of the Çingeneler (Gypsies) by the majority (non-Roma) to find that the Roma and their identity is almost unilaterally rejected. They are subject to a "silent" and institutional discrimination, a prejudice that is made overt in the a range of Ordinances and Laws that have been instituted since the beginning of the modern Turkish state, laws which call for the assimilation or "ejection" of Roma from various regions and cities.

A small, but increasing number of Turkish Roma are aware of the international Romani movement and its struggle for the rights and self-identity of the Roma throughout Europe and the rest of the world. The number of these Roma who are as yet politically "activated" or "aware" is still very small but the situation is slowly improving. Their increased involvement in sustainable development programmes is a progressive step into the future as it is only through awareness-raising, enhanced community organisation and increased legal/societal representation that social change will be stimulated and Roma access to social and educational resources will be improved.

References

Andrews, P. (1992) *Turkiye`de etnik gruplar*, (Trans: Mustafa Kupusoglu), Istanbul: Ant Yayinlari, Tumzamanlaryayincilik

Arayici, A (1999) *Ulkesiz bir alk. Cingeneler*, Istanbul: Ceylan Yayinlari,

Beuninghaus, R. "Les Tsiganes de la Turquie Orientale" in *Etudes Tsiganes*, 1999, Vol. 1.

Marushiakova, E. and Popov, V. (2000) *The Gypsies in the Ottoman Empire*, UK: University of Hertfordshire Press (Interface Collection)
Marushiakova, E. and Popov, V. (1997) *Gypsies (Roma) in Bulgaria* Germany: Peter Lang Verlag

CHAPTER SEVEN

SOME THOUGHTS ON THE SOCIAL HISTORY OF THE CZECH ROMA

ALAN OWENS

This essay seeks to provide a brief overview of some aspects of the historical experience of those communities often referred to as Roma in Central and Eastern Europe. The origins, background, and historical development of these communities will be examined on a broad scale. Following from this will be a closer examination or (brief) case study, of the history and experiences of the Roma in the territories of the former Czechoslovakia. This will focus on how this group were treated under the totalitarian regime, with a particular focus on the process of sterilisation of Roma women that took place between 1970 and 1990. Finally, the essay will conclude with a brief examination of how the experiences of the Roma in Europe compare with the experiences of other minority groups in sedentary societies, namely Travellers in the UK and Ireland. This will look at whether such things as sterilisations took place among these groups, and how discrimination and policies of assimilation are a recurring theme among all of these groups.

It is very difficult to present a general history of the Roma, because there are many views on what form such a history might take. This aside, it is also difficult to present a unified history of all Roma, because of different historical migrations, the dispersal of different tribes, and the fact that many groups of Roma were not, in fact, Nomadic and have long been assimilated into the population or culture of whatever country they chose to settle in. Indeed, even the term Roma is a problematic one, as this group, that is believed to number somewhere in the region of 15 million people in Europe, has been referred to under many names, *Gypsies, Tsigani*, *Tzigane*, *Cigano*, *Zigeuner.*[1] Most would use the term Roma or

[1] The Patrin Web Journal, A Brief History of the Roma, http://www.geocities.com/~patrin/history.htm

Rom, or their tribal names, which we will examine shortly. For the most part of their history, Roma have been referred to as Gypsies, and here both terms are intended to be interchangeable.

However, these difficulties noted, there are a number of studies that trace the origins and history of the Roma in general terms. This is something I will present, followed by a closer, detailed study of Roma in the territories of the former Czechoslovakia. The principle reason for there not being a complete, definitive history of the Roma is the relative lack of any written sources documenting the early origins of this group. It is fitting then that the Primary evidence regarding these origins should be linguistic in nature, as Romani has for a long time been the 'language of an unlettered people'.[2] Indeed, as one scholar noted, 'The true history of the Gypsy race is on the study of their language'.[3]

It was first suggested about 200 years ago that Romani was of Indian origin, as emerging linguists noticed similarities between the vocabulary of Romani and some Indian languages.[4] The official view is now that Romani is of Indo-Aryan origin, with its roots in Punjab, or Hindi.[5] Scholars in the 18th Century recognised that basic words, such as those for numerals, or body parts were particularly identifiable as Indian. Angus Fraser has done some work on this, adopting written Romani language conventions that were implemented by the Romany World Congress in 1990, and comparing these to Sanskrit and Hindi. He notes that the resemblances between Romani and Sanskrit or Hindi in these basic vocabulary terms are clear, and that the Romani dialects he chose to examine contain at least 500 words which are recognisable as being of Indian origin.[6]

Some studies have shown that it is not just linguistic evidence that show these origins. Others point to similarity of customs, of social structures and profession choices.[7] Fraser indeed examines Physical Anthropology, blood group studies and population genetics. However, he notes that these are suggestive rather than conclusive.[8] Studies like this and others like it, have produced a generally accepted consensus that the Roma originated from Northern India, and may have begun to spread into

---

[2] Angus Fraser, *The Gypsies,* (Oxford, 1992) p.13
[3] Ibid, p.10
[4] Ibid, p.14
[5] Patrin, A brief history, http://www.geocities.com/~patrin/history.htm
[6] Fraser, *The Gypsies,* p.17
[7] The History and Origin of the Roma, http://romove.radio.cz/en/article/18158
[8] Fraser, *The Gypsies,* p.25

Asia in roughly 300 BC.[9] Scholars believe a number of migrations may have taken place, because of the upheavals of wars of the time, and indeed some Macedonian Roma folk legends trace their roots back to Alexander the Great in the fourth century B.C.[10]

As stated above, some scholars believe that a number of 'great' migrations took place. The first was the initial migration from Northern India to Asia, followed by a second from southwest Asia into Europe in the Middle Ages.[11] However, while some sources note references to the presence of Gypsies in the 11th Century in Eastern Europe, it is not until the 12th century that substantiated historical documents note the presence of Roma in Europe.[12] These include the records of the Republic of Ragusa, now the Croatian city of Dubrovnik, which suggest a rich Gypsy life in the Republic, and Hungarian and Slovak records, which feature the first mention of Roma in Hungarian accounts.[13] These records also note the growing movement of Roma through Central Europe over the next number of decades.

Many Roma in Europe at this time (c.12th-15th Century) were free and many were employed as metal craftsmen, musicians and soldiers. Hungarian and Slovak records detail how many Gypsy smiths and soldiers were part of their military forces, and Lithuanian records detail how they were viewed as 'useful citizens'.[14] However documents increasingly refer to trade involving Gypsy slaves, and attitudes began to change due to the growing Ottoman threat, and the Reformation.[15] Records show Roma began to increasingly arrive in Poland in the 16th Century, fleeing harassment, and gradually began to spread into Russia.[16] There is evidence to suggest that Roma and Gypsies began to be viewed as Turkish spies in the territories of Hungary and Slovakia, and that subsequent attempts to regulate them pushed them towards a more nomadic way of life.[17]

---

[9] Fraser, *The Gypsies*, p.22

[10] David M. Crowe, *A History of the Gypsies of Eastern Europe and Russia,* (New York, 1994), p.xi

[11] A third migration is believed to have been from Europe to the Americas in the 19th and 20th Centuries, and a fourth following the fall of the Iron Curtain. From The Patrin Web Journal, *A Brief History of the Roma,* http://www.geocities.com/~patrin/history.htm

[12] Crowe, *A History*, p.xi

[13] Ibid, p.xi

[14] Ibid, p.xi

[15] Ibid, p.xi

[16] Donald Kenrick, *Gypsies: From the Ganges to the Thames,* (Hertfordshire, 2004), p.

[17] Crowe, *A History of the Gypsies of Eastern Europe and Russia,* p.xi

These attempts at control can be seen repeatedly in Roma history, as well as the existence of an active Gypsy slave trade in the 16th and 17th Centuries. Austrian-Habsburg policies of the time were similar to those used in Eastern Europe and Russia over the next two centuries.[18] These policies attempted to end the pattern of nomadism, and involved such things as placing Roma children in foster homes, a policy that has contemporary relation, as we shall see. Roma began to arrive in Russia in significant numbers in the 17th and 18th centuries, and while they had an initial and occasional impact on culture, generally continued to be relegated to the lowest class in society. In fact, despite the considerable contributions of Roma to culture in Eastern Europe and Russia, in music, literature and theatre, this did not particularly serve to earn them much respect from society.[19]

The upheavals of the 19th and 20th Centuries are well-documented, as are the horrors of the 'Pořajmos' or Gypsy Holocaust, that saw the deaths of hundreds of thousands, indeed millions of Roma and Gypsies. These events are better explored in another case study, but must be mentioned in any examination of Roma history. The following Soviet takeover of Eastern Europe did little to improve the situation of the Roma, as strong ethnic identity was discouraged, and assimilation encouraged.[20] An example of the treatment of Roma in this totalitarian system will shortly be examined, as we move the focus of this study to a narrow look at the history and experiences of Roma in the territories of the former Czechoslovakia.

The reason I chose this area to examine in particular was my discovery on the ERRC (European Roma Rights Centre) website of a report detailing the coercive sterilisations of Romani women in the Czech Republic and former Czechoslovakia. In the course of my research I also discovered a contact in England that believes similar practices have taken place on travellers in the UK. But before we discuss these issues, we examine the history of the Roma in Czechoslovakia.

First it must be highlighted that the territories of modern day Czechoslovakia were made up of the Czech lands of Bohemia and Moravia, Silesia and Lusatia, and the Slovak portion of the Hungarian kingdom. The Roma entered these lands at some time in the early Middle Ages. Initially, it seems, large travelling groups of Gypsies were well received, carrying letters of safe conduct from King Sigismund of Hungary, and the 'nobility of the leaders' of these groups was noted in

[18] Ibid, p.xiii
[19] Ibid, p.xiii
[20] Ibid, p.xv

official Slovakian records.[21] This period of initial acceptance is similar to events in other parts of Europe, in that it was too followed then by increasing distrust of these groups. Again this is somewhat linked to the rise of the Ottoman Empire, and the view of Gypsies as being Turkish spies or mercenaries, and the outbreak of fires in Prague in 1541, was blamed on Gypsies.[22] This was exacerbated by the view of the Church that Roma were 'indifferent to religion' and Roma became targets of persecution in these lands.[23]

Following the fires King Ferdinand I ordered the expulsion of Roma from Czech territories. In the 17th Century Gypsies were hanged from trees along the border to dissuade others from entering, while in Slovakia it was decreed they would be spared this if they settled after three weeks.[24] Policies of persecution were ended in the 18th Century by the Austro-Hungarian Empress Maria Theresa who wished to assimilate Roma into society, by forbidding the policy of nomadism.[25] These policies, and those continued by her contemporaries and successors, were often quite harsh and brutal. Despite these policies the Roma continued to resist assimilation and sought to maintain their cultural identity throughout the 19th Century. This century was one of wars and upheavals that continued to produce appalling conditions for Roma that stayed in these territories. Conditions of poverty, suppression of education and employment discrimination among Roma became a common occurrence through this century and the next and little changed with the establishment of the Czechoslovak Republic in 1918-1919. Gypsies were recognised as a separate nationality in 1921, but this did little to change widespread opinion that they were a burden on society.[26]

This is not to say that all Roma or Gypsies were subjected to appalling conditions and policies in this region. There is evidence to prove that a large number of Roma settled peacefully in Slovakia, and while they did not make much advance up the socio-economic ladder, the 1921 census

---

[21] These records are the Book of Executions of the Lords of Ruzomberok. From Crowe, *A Brief History,* p.33

[22] David M. Crowe and John Kolsti, *The Gypsies of Eastern Europe,*(New York, 1991), p.94

[23] The history of the Roma minority in the Czech Republic, http://romove.radio.cz/en/article/18913

[24] Helsinki Watch report, *Struggling for ethnic identity, Czechoslovakia's endangered Gypsies,* (USA, 1992) p.5

[25] Crowe, *A History of the Gypsies of Eastern Europe and Russia*, p.38

[26] Crowe and Kolsti, *The Gypsies,* p.95

shows 7,967 Roma living in Slovakia.[27] However, in general conditions were not good for the majority of Roma, and would only worsen in the interwar period, and finally, descend into horror during WWII. Most Roma in the Czech lands of the Third Reich, along with German Roma perished during these times. This corresponded with the fact that Czechoslovakia ceased to exist as an entity during WWII. However, in the Slovak Republic, which was also governed by a fascist regime, things were not as severe. Here, the more numerous Gypsies survived, and while forced to endure extreme discrimination and exclusion, were not exterminated.[28]

Following the war, with the 'Czech' Roma almost entirely wiped out, the Czech lands were repopulated by Roma primarily from Slovakia, but also from Romania and Hungary.[29] For the first time, the Government encouraged (forcefully at times) Gypsies to come and seek employment in the new Czechoslovakia. This was the first time that Gypsies entered the market as unskilled workers, particularly in urban areas, but they still lived separate lives, congregated in ghettos, and were not assimilated into society.[30] However, it is said that 'Their labour served, in part, to put the Czechoslovak economy back on its feet'.[31] In Slovakia, Gypsies began to move to and from Czech territories and the size of the Gypsy population began to swell. All of these settlements were characterised by overcrowding and unsanitary conditions, and the Czecholoslovak Socialist Government's response was to launch a campaign to fully integrate the Roma into the community and the workforce.[32]

This was in line with the USSR banning nomadism, a policy that was implemented by most Eastern European Communist states. This policy, and the one implemented by the Czech regime have echoes in the past with the policies of Maria Theresa. In order to quell strong ethnic identity, Romani identity was discouraged as a 'relic of the past' and other policies such as strategies of population dispersal (Law 74), which formally deprived thousands of Roma their right to freedom of movement, were

---

[27] Crowe, *A History of the Gypsies of Eastern Europe and Russia*, p.44. Census figures from the period vary widely, from figures of 31,415 Roma in the 1930 national census, while a Slovak Rom census from two years earlier suggests figures of 60,315.

[28] Crowe and Kolsti, *The Gypsies,* p.96

[29] Will Guy, *Between Past and Future, the Roma of Central and Eastern Europe,* (Hertfordshire, 2001) p.38

[30] Crowe and Kolsti, *The Gypsies,* p.96

[31] Will Guy, *Between Past and Future,* p.289

[32] Will Guy, *Between Past and Future*, p.290

introduced in 1958.[33] This campaign failed for several reasons, Czech employers ignoring regulations, Slovak authorities were reluctant to house Gypsies and thus disperse them outside their settlements, and Czech local authorities refused to accept larger scale relocations of Roma from Slovakia.[34] Things did not improve much over the years, a 1970 Eastern Slovak report documented how the plan to eliminate settlements could not be implemented because 70,000 Gypsies still lived in shanties, where 'each adequate well and toilet had to serve over two hundred Gypsies'.[35] Indeed this situation and others like it throughout Czechoslovakia saw constant deterioration due to the combination of appalling conditions, the high birth rate and subsequent steady increase in population.

It was this situation that led to a policy that is maybe the most shocking of all directed toward the Roma throughout the ages, but more so because of its contemporary nature. This was the one of 'Population control' which took a specific form in the case of the Roma. A dissident group representing Roma documented this and other human rights abuses alleged against the Communist government. Charter 77 maintained that the denial of minority status for Roma was 'dictated by the desire of the ruling powers to reduce the size of the minority problem'.[36] This population control took the form of 'coercive sterilisations'. This practice took the form of sterilising Romani women either under duress, or by offering financial incentives, or simply without the knowledge of the women during a caesarean section or abortion.[37]

This process was documented and investigated by the Helsinki Human Rights Watch. They found that Roma women were often sterilised without their consent, or were pressured into agreeing by government social workers and health workers. Most however, reported that they were enticed by monetary incentives offered to them by social workers.[38] The report documents how the government allegedly took specific steps to encourage the sterilisation of Roma women in order to reduce the high proportion of the population they made up, and as a result, a disproportionately high number of Roma women were indeed sterilised, often in violation of the safeguards of the Public Decree on Sterilisation.[39]

---

[33] Ibid, p.291
[34] Crowe and Kolsti, *The Gypsies,* p.99
[35] Ibid, p.99
[36] Will Guy, *Between Past and Future,* p.292
[37] Helsinki Watch Report, *Struggling for ethnic identity,* p.19
[38] Ibid, p.19
[39] Ibid, p.19

Government reports obtained by the Human Rights Watch reflect that the Government viewed this as an acceptable and necessary policy. The report shows the awareness that monetary incentives would encourage women to accept sterilisations, that doctors and surgeons reflected state policy in dong this whether voluntarily or involuntarily, and even shows how social workers reflected the Government's belief that Roma communities were 'inbreeding to an unhealthy degree'.[40] It concludes that the practice ended in the mid-90's but not before monetary incentives continued to be offered for sterilisations, and despite which many doctors involved continued to practice medicine in 1992.

This practice compares worryingly with practices that may have, according to one source I discovered, taken place in the UK, among Gypsy and Traveller women. This source, who wished to remain anonymous, has personal experience of this. His own wife was rushed into a sterilisation will still in hospital after giving birth, with no consultation with him. The couple noticed that many of their 'contemporaries' had also stopped having children, and a few were prepared to discuss their own experience of 'rushed' sterilisations. This source also claims to have a local council document that relates to these sterilisations and states an aim 'to eliminate those who are capable of recreating the species'.[41]

This practice has possibly taken place in Ireland among Traveller women, but no evidence exists to confirm this as yet. Certainly comparisons can be drawn among these minority groups, be it Travellers in Ireland or the UK or Roma in Eastern Europe; all have faced policies of exclusion, persecution and forced assimilation. Even in Ireland, the State has always lent towards policies of assimilation, as the 1960 Commission on Itinerancy shows, which had far reaching implications for public policy. It equated itinerancy with poverty, took no heed of cultural differences, and approved of assimilation as the answer to a problem that has no 'Final Solution'.[42]

Practices such as coercive sterilisation, while at the extreme end of the scale, are also indicative of practices and policies of persecution, segregation, rights deprivations, human rights abuses, exclusion from education and employment that took place in Czechoslovakia and other countries over the history of the Roma. Some, if not all of these abuses are now documented by groups such as the ERRC, and are recognised by the

---

[40] Ibid, p.27

[41] From Email correspondence with a source involved in Gypsy and Traveller Law Reform in the UK. Name withheld.

[42] From Lecture notes, HI6052: Irish Travellers, Roma and Migration in Ireland 1850-1980, Micheal O'Haodha, Week 6, University of Limerick.

European Union, who have attempted to improve the situation of Roma in Europe.

So where does the future of Roma minorities in Central and Eastern Europe lie? With the development of democracy in former totalitarian states, the Roma have been presented with a choice that has its roots deep in history. There has been a huge growth in groups and governments working with and for Roma minorities. Thus the Roma can choose to assimilate and abandon their culture and traditions, or they can choose to define themselves by this culture and lifestyle.[43] While the latter path is certainly a more difficult one, it is one that is worth taking as the Roma seek a better future for themselves than the one that a glimpse at their past reveals.

References

Barany, Z, (2002) *The East European Gypsies, Regime change, marginality and ethnopolitics,* Cambridge: CUP

Crowe, D.M, (1994) *A History of the Gypsies of Eastern Europe and Russia;* New York: I.B. Tauris

Crowe, D.M and Kolsti, J (eds.), (1991) *The Gypsies of Eastern Europe;* New York: M.E. Sharpe

Danbakli, M (ed.), (2001) *Roma, Gypsies: Texts issued by International Institutions;* Hertfordshire: University of Hertfordshire Press

Fraser, A. (1992) *The Gypsies,* (Oxford: Oxford University Press

Guy, W. (ed.) (2001)*Between Past and Future, the Roma of Central and Eastern Europe;* Hertfordshire: University of Hertfordshire Press

Helsinki Human Rights Watch (1992) *Struggling for Ethnic Identity, Czechoslovakia's Endangered Gypsies,* USA: Helsinki Human Rights Watch

Kenrick D. (2004) *Gypsies: From the Ganges to the Thames;* Hertfordshire: University of Hertfordshire Press

Websites Referenced

www.errc.org
http://www.romani.org
http://www.geocities.com/~patrin/
http://www.romahistory.com/en/1-1.htm
http://hrw.org/reports/1996/Czech.htm

[43] Crowe, *A History of the Gypsies of eastern Europe and Russia,* p.xvi

CHAPTER EIGHT

THE ELITE MEDIATION OF PREJUDICE AGAINST BRITAIN'S PAVEES: A CASE OF "REPROBATE" MOBS, SUCCESSFUL ASSIMILATION, OR DISQUALIFICATION?

COLM POWER

Introduction

This essay links a number of issues concerning the lived experience of Pavees (generally known to sedentary people as Irish Travellers) in Britain and is based on some of the findings of a major recent research project into the social situation of Irish Travellers in Britain (Power, 2004). It begins by explaining who Irish Travellers are from an ethnic perspective and historically how they came to be an indigenous ethnic group that now has legal recognition as such in Britain. It also describes changes over time in the nomadic Traveller economies that led to increased migration into Britain from Ireland and how Irish Travellers have struggled to adapt to rapidly changing socio-economic, political and legal contexts since the 1950s. The difficulties and barriers to ascertaining accurate demographic profiles and some of the consequences of these are also discussed, followed by a section briefly describing the remit for the research project and report on which this piece is based. This is followed by brief summaries and discussions of the few previous bespoke research reports into the social situation of Irish Travellers in Britain and a short discussion of the most vulnerable sections of this community.

The next section examines the impact of specific and general legislation on the lives of Irish Travellers in Britain and the move from 'laissez faire' to government legislation and policies aimed essentially at the assimilation of indigenous nomadic groups. Accommodation policies were often reluctantly put into practice and tended to exclude Irish

Travellers as they were not viewed as indigenous by many local authorities. The socio-political context of Thatcherism from the mid 1980s to mid 1990s prompted a move from policies of assimilation to coercive state legislation that criminalised many aspects of Irish Traveller culture. But the advent of a New Labour government in 1997 with a strong commitment to the social inclusion of minorities promised an improvement in the lot of Irish Travellers in Britain. Contrary to sanguine expectations of improvements in the situation none of the coercive laws introduced by the Tories have been withdrawn, while an examination of the discourse in relation to Irish Travellers employed by a number of key government ministers indicates that Irish Travellers still remain outside the limits of liberal pluralism in Britain today.

Who are Britain's Irish Travellers?

This brief historical perspective outlines recent Irish Traveller migration to Britain and explains the social, legal and administrative contexts that impact on Travellers' lives. Irish Travellers usually refer to themselves as 'Pavees'. They are an indigenous nomadic minority group both in Ireland and Britain.[1] They have been part of Irish and British society for many centuries according to historical sources (McDonagh and McVeigh, 1996; Fraser, 1992). Irish Travellers' distinctive way of life, values, culture and traditions manifest themselves in Traveller 'nomadism', the centrality of the extended family, their own language or 'secret patois', a strong sense of the spiritual dimensions of life, and the entrepreneurial nature of their economy (Ó Riain, 1997; LIWC, 1995; McCann et al, 1994; Power 2004). Irish Travellers, and indeed other Traveller groups, do not form discrete or homogenous ethnic categories in the sense that suit sedentary institutions and bureaucracies that tend to compartmentalise in order to generalise, but they can be better understood from without as *'a community of communities'* (Parekh, 2000: 34). The extended family is the central Irish Traveller organising social unit. These families are linked to bigger clans, while alliances (or disputes) between or within particular clan groups form the bonds or demarcations that link (and divide) these groups in an 'ethnographic' sense. Many Irish Travellers still speak their language, known as 'Gammon' or 'Cant', to some degree though it is now in decline (Cauley, 2005). The language is a combination of 'disguised' Irish and English words that uses English

---

[1] They also have a number of recognised long-term settlements in the United States.

language syntax and sentence structure (Binchy, 1994). Irish Travellers use Cant/Gammon in conjunction with English and were very reluctant to divulge it to outsiders historically. Travellers use the language within their own community and in the presence of settled people in pressured situations (ibid).

Travellers have also played a significant role in maintaining aspects of Irish culture like music and storytelling, and have helped to reinvigorate aspects of the struggling folk music tradition in Britain. Rural settled people often welcomed Irish Travellers as a source of news, entertainment, consumer goods and as flexible seasonal agricultural workers before the advent of mass media, consumerism and telecommunications. They filled vital niches in the economy of pre-industrialised Ireland and parts of rural Britain up to the 1950s as they bartered, sold and recycled scarce consumer commodities. Historically, Irish Travellers have been employed in horse trading, seasonal farm-work, rural crafts, selling domestic goods door-to-door (hawking), and as tinsmiths and tradesmen. Many Traveller extended families had established regular migratory patterns that often stretched beyond national boundaries to sustain their lifestyle. One example of an Irish Traveller family (consisting of about thirty members) migratory pattern in the early 1950s was to cross from Ireland to north Wales for Spring lambing, then travel and work through the English Midlands to eventually reap the harvest in rural Lancashire. They would then return to the north of Ireland to spend some time with relatives and finally winter in Wicklow in the southeast before the cycle began again (Irish Traveller woman cited in Power, 2004: 5).

Urbanisation, mass production of cheap disposable plastics and other domestic items, and the mechanisation of agriculture changed the lives of Irish Travellers profoundly and undermined the basis of their rural economy in Britain and Ireland since the 1960s. Travellers migrated in large numbers to urban areas such as Dublin and the major cities in Britain (replicating the migratory patterns of settled rural Irish migrants) for employment opportunities from the late 1950s onwards (Gmelch and Gmelch, 1985; Power, 2004). The common feature of individual Irish Traveller economies is nomadism and self-employment. A combination of social separateness, nomadism, cohesive extended families, and a strong resistance to wage-labour underpin the nature and structure of the Traveller economy, differentiating it significantly from most of the settled community (Okely, 1994). Traveller's entrepreneurial skills adapted to new employment opportunities as the 'mainstream' economy changed and developed. Access to motor vehicles gave rise to new opportunities for commercial nomadism (ibid, 1980). This pattern is corroborated by a

1998 study in Ireland which estimates that up to twenty percent of the approximately 15,000 market traders in Ireland are Irish Travellers (McCarthy and McCarthy, 1998: 51).

As a consequence Traveller employment patterns have shifted to casual forms of building work, tarmacing, tree-topping, market stalls, gardening and scrap metal collection.[2] A minority have developed very successful businesses dealing in antiques, furniture, carpets and linoleum. Street begging and petty crime amongst Travellers is a minor concern, but Travellers as a whole have been routinely labelled as petty criminals by some settled people and sections of the media (Rutter, 1997; Acton, 1994). Anti-Traveller racism and discrimination increased dangerously in Ireland after urbanisation and industrialisation in the 1960s and the breakdown of the old rural hierarchical, but also mutually beneficial, relationships between settled and Traveller. Migration to the periphery of major cities in Britain has increased in the last half century as Travellers traditional economies have been modified by modernisation. Their nomadic rights have been severely curtailed by criminal justice legislation, commodification of marginal land, and settled people's resistance to their nomadic way of life. Conflict has arisen between urban settled denizens, municipal authorities, police forces, and Irish Travellers over modes of living and access to scarce resources (Sibley, 1995).

Irish Traveller demographics in Britain

Gmelch and Gmelch (1985) interpreted the migration of Irish Travellers into Britain as an adaptation strategy, driven by the processes of rapid urbanisation, agricultural mechanisation, and industrialisation in Ireland from the 1960s onwards. Travellers had already been leaving Ireland, at least since the Great Famine in the 1840s up to the 1950s, due to diminishing economic opportunities in an increasingly depopulated and contracting rural Irish economy. Disparities in estimates of Traveller populations are usual across European states and often reflect negative attitudes of states and settled populations to ideologically disqualified, politically invisible, and socially marginalised nomadic groups (Clark, 1998). The large disparities in official population estimates can also be partly explained by high internal mobility rates, but also by the (largely un-researched) phenomenon of Irish Traveller seasonal migration within and between Britain and Ireland (north and south), continental Europe, and even the United States.

[2] This contributes greatly to recycling of scrap metals and other materials.

Accurate statistics on Irish Travellers are difficult to obtain as they are not yet specifically defined as a separate ethnic grouping on Census forms in Britain.[3] Travellers and Gypsies who are deemed to be living in 'settled' accommodation are not enumerated in any way by the state, so that the inability of many Irish Travellers to live a traditional nomadic lifestyle due to anti-nomadic legislation and practices leads directly to their ethnic disqualification. Current Traveller population estimates by the Office of the Deputy Prime Minister (ODPM)[4] are based on local counts of trailers on official sites or other 'known' unauthorised encampments carried out by local authorities. The 'total population' is extrapolated from the 'estimated' occupancy per trailer. This reporting procedure is non-mandatory and incomplete – no information is collected on specific Gypsy and Traveller population numbers, age, gender, or ethnicity - as a result no accurate government statistics exist for Irish Travellers in Britain (Hickman and Walter, 1997: 20-21). Recent government figures for England (ODPM, July 2003, www.housing.odpm.gov.uk/information/index14.htm) indicate that 14,736 trailers were counted by local authorities overall.[5] Of those, 3,976 were in unauthorised camps, indicating that just over twenty-four percent of all 'counted' Travellers and Gypsies have no legal stopping places and are constantly under threat of harassment and eviction with no secure place to reside.

Many Irish Travellers are now seasonally nomadic, preferring to winter in housing while risking the vagaries of travelling for work during the Summer months (Irish Travellers cited in Power, 2004). The above ODPM estimates do not include the large numbers of Travellers who are living either temporarily or permanently in settled forms of accommodation. Numerous 'settled' Irish Travellers have been forced from their traditional nomadic way of life due to lack of trailer sites and the virtual outlawing of Traveller nomadism in England and Wales initially by "Section 39" of the Public Order Act (1986) and later by the extended powers contained in the Criminal Justice and Public Order Act (CJPOA 1994). No official data are collected on 'settled' Travellers, but these supposedly assimilated Travellers still 'travel' through private, association and LA housing stocks, often moving between or within cities up to five or six times a year (Power, 2004). Large disparities in

[3] Both Census 2002 in Republic of Ireland and Census 2001 in Northern Ireland had specific categories for Irish Travellers.

[4] Formerly the Department of Environment, Transport and the Regions (DETR).

[5] These figures include all indigenous nomadic groups and are not specific to Irish Travellers.

population estimates are partly explained by high internal mobility rates, but also by the phenomenon of Irish Traveller international seasonal migration between Britain, Ireland and beyond. Both international seasonal and more permanent migration is not ethnically monitored, but research interview evidence from Irish Travellers indicates complex patterns of migration related to individual Traveller families' responses to modern economic opportunities (Irish Travellers cited in Ibid).

Since the CJPOA 1994 removed the statutory duty on local authorities to provide Traveller sites, very few new sites are being built to compensate for site closures and Traveller population growth.[6] Population estimates for all Travellers and Gypsies in the United Kingdom range between 90,000 and 120,000 (Niner, 2002: 10). The age profile of Irish Travellers is exceptionally young: they tend to marry early and families average eight children, therefore the population is growing quickly (ibid, 2002: 9-10). Irish Travellers form a large proportion of the nomadic and semi-nomadic Traveller population in Britain. Recent research estimates that about 15,000 nomadic Irish Travellers live in Britain (O'Dwyer, 1997: 9). Again this estimate does not include 'housed' Travellers - often inaccurately assumed to be assimilated. Irish Travellers interviewed for the 'Room to Roam' project and then living in settled accommodation before the British Census in June 2001 all indicated that they would not return any Census forms - this was due mainly to their distrust of all official information gathering exercises (cited in Power, 2004). Irish Travellers are often institutionally and spatially invisible as a distinct ethnic community in Britain, relegated to 'defiled' and neglected inner city or urban peripheral areas whether sited or housed (Sibley, 1995: 72-88). The irony for Irish Travellers is that while they are an officially recognised ethnic group since August 2000 in England and Wales (Discrimination Law Association, 2002: 10) for the purposes of the Race Relations Act (RRA, 1976) and the Race Relations Amendment Act (RRAA, 2001), their nomadism and much of their traditional way of life is still criminalised.

Research background

This essay is based mainly on primary data gathered for 'Room to Roam: England's Irish Travellers' report - a major Community Fund resourced social research project (see Power, 2004).[7] This three-year

[6] Local authority sites lost a total of 234 trailer pitches in Britain from 1998-1999.

[7] This project was run by a consortium that included Action Group for Irish Youth, BIAS Irish Travellers Project, and the University of Surrey - St. Mary's College.

project involved qualitative research including focus groups and 140 in-depth interviews with Irish Travellers and relevant service providers (health, education, social services, politicians, voluntary services, and criminal justice system) in two major English cities – Manchester and London. The criminal justice element of the research project was enhanced after the project began as a result of growing research evidence suggesting relatively high numbers of Irish Travellers in the Criminal Justice System who were experiencing discrimination (Power, 2003b and 2004). The research aimed to assess the impact of social marginalisation, ethnic disqualification, and criminalisation on Irish Travellers' health and life-chances in England. This research provided the statutory and voluntary agencies involved with Irish Travellers' health, education, social welfare, and criminal justice provision, with locally specific information that could be mobilised for culturally sensitive policy development, service provision and delivery throughout England and beyond. Power's (2004) work endeavoured to inform Traveller policies generally at a national level – for example, the report has been adopted as a core source by the British government's Social Exclusion Unit and has influenced policy and practice changes in the English Probation Service's work with Travellers.

Irish Travellers are discriminated against and excluded both institutionally and socially while their status as a distinct ethnic group in Britain is barely recognised. As such they have received scant attention from government, academic social researchers, policy formulators, or statutory service providers – this has burdened them with distinct disadvantages in accessing healthcare, educational and training support, and general social welfare. The impact of anti-nomadic legislation, covert assimilation policies and modernisation stresses have resulted in the partial decomposition of Irish Traveller cultural tenets, extended family structures and self-esteem leading to the criminalisation of large sections of this community. The 'Room to Roam' research identified and evaluated some of the causes of poor health, inadequate social care, institutional discrimination, educational disadvantage and criminalisation of Irish Traveller communities. It explored ways of developing and providing more reflective and culturally sensitive institutional policy strategies and resultant practices in key areas of service delivery that could promote Irish Traveller empowerment (Power, 2004).

The remit of 'Room to Roam' was to critique, synthesise, develop, and re-interpret distinct and fragmented bodies of knowledge regarding Irish Travellers in Britain coupled with a major qualitative primary data collection exercise. Previous research work frequently reflected the piecemeal development of Irish Traveller social policy and practice by institutional agencies inappropriate to the specific cultural disposition of this distinct ethnic grouping. Research carried out by voluntary bodies or local authorities often betrayed a localised and disjunctive focus due to the vicissitudes of funding research into marginal or 'invisible' ethnic groups. Power's (2004) work draws attention to the contingent, shifting, and multiple boundaries that generate individual, collective, and institutional practices of social exclusion and inclusion. Power's (2004) report explores ways of developing and providing more reflective and culturally sensitive institutional policy and practice in key areas of service delivery, including strategies that promote Irish Traveller empowerment and influence on the design and delivery of key services. This research report, unique in its comprehensive treatment of the needs of Irish Travellers in England, is designed to involve and positively inform Irish Travellers and other similar communities throughout Britain and beyond.

Select review of literature on Irish Travellers in Britain

A substantial body of research about Irish Travellers has been conducted and published in both the Republic of Ireland and Northern Ireland over the past twenty years. This research has invariably raised serious concerns about the very high levels of poverty and disadvantage experienced by this community and has underlined deep concerns around the very poor health profiles, mortality rates, lack of suitable accommodation, social exclusion, and lack of provision and access to appropriate education, welfare and support services for Irish Travellers (Morris and Clements, 2001; Power, 2004). The research reviewed below is chosen because it is directly relevant to Irish Travellers experiences in Britain.

Little detailed research has been carried out in England into the particular welfare needs of Irish Travellers. Much of the research carried out has examined Gypsies and Travellers' needs in a 'general' sense without specific reference to the particular disposition or needs of Irish Travellers as a distinct ethnic group, or have treated Irish Travellers as an addendum to other indigenous nomadic groups or as a 'sub-group' of the settled Irish population in Britain. The 'Room to Roam' research project attempted to treat Irish Travellers as a distinct ethnic group in their own right. The following paragraphs review some of the research undertaken over the last twenty years that concentrated on Irish Travellers in Britain.[8]

A 'London Strategic Policy Unit' report compiled by Debra L'Esteve (1986), a Policy Advisor on welfare benefits to 'The Travellers and Welfare Benefits Working Party' set up in June 1986 because of the serious concerns of Education Welfare Officers, Health Visitors and the Child Poverty Action Group about the treatment of Travellers, outlines high illiteracy rates, the racist attitudes of some benefit staff and related poor access to social security benefits by Travellers. As of 1985 Circular S50/85 required the Department of Health and Social Security (DHSS) to demand original birth certificates, National Insurance numbers and wage slips which Travellers often do not possess. These guidelines were issued because of assumptions made by the DHSS that Travellers were committing fraud, though no evidence was produced to back these claims. This heralded a shift in emphasis from service delivery to the investigation of fraud by the DHSS. This policy was taken to extremes by the North London DHSS who issued Identity (ID) cards to be filled out by their staff when dealing with Travellers that required information such as hair colour, height, darkness/lightness of skin, and whether the claimant has a known criminal record.

A 'Nomadic Claims Working Group' was set up by the DHSS which involved developing a regional/national computerised record of claimants including physical appearance etc. During any fraud investigation, Travellers could receive 'Urgent Needs Payments', but these are lower than supplementary benefits. Travellers are rarely informed by DHSS staff of other benefits which they are entitled to such as 'Attendance Allowance'. The take up of 'Child Benefit' is proportionately much lower amongst Travellers and when supplementary benefit is assessed it is often assumed that Travellers have already claimed/received 'Child Benefit', therefore many Travellers lose out substantially. Camden Citizens' Advice Bureau brought the existence of the DHSS ID card system to light, and challenged it. Wood Green Citizens' Advice Bureau also challenged this ID card practice and eventually it was agreed that Travellers could pick up their claims from Wood Green if copies of birth certificates were obtained through the Citizens' Advice Bureau. The report concludes that even when Wood Green Citizens' Advice Bureau had developed good working relations with Travellers around benefits issues, police actions

[8] These reports concentrate on urbanised and 'sited' Travellers who happen to be in the main Irish Travellers in these particular reports.

undermined their work. A police raid resulted in all Travellers on a site over the age of eight being arrested and fingerprinted. Some benefits were stopped after this raid with no reasons given by the DHSS.

Hyman (1989) emphasised disparity in access to healthcare between those living on authorised, legal sites and those on unauthorised encampments; the ability to register with a GP was much higher for those on permanent sites. She stressed the importance of outreach work by health visitors for those without a legal site and remarks on the reluctance of GPs to make home visits to unauthorised sites.

'Rights for Travellers' (1995) produced by the London Irish Women's Centre (LIWC) surveyed local authority provision for Irish Travellers in London. The report revealed that the Irish Traveller population had increased substantially in recent decades despite the CJPO Act 1994. Approximately 3,200 Traveller families (mainly Irish) in London did not have access to an authorised site in 1992, forcing more Irish Travellers to move into housing. The report points out that Irish Traveller children are more at risk of being taken into care and reported on a social worker's visit to a juvenile detention centre in 1990 that found eleven percent of the boys detained were Travellers. The report criticises popular assumptions that Irish Travellers are somehow getting 'a better deal' than settled people. This survey was distributed to thirty-three London local authorities and twenty-two responded. Most local authorities had no standardised method of collecting data on Travellers. The report criticised the lack of security for Travellers' tenure on sites. Travellers pay rent and Council Tax but do not have equal rights with 'settled' tenants. In 1994 the highest rent charged (for a permanent licensed site) was in Hackney at £55 per week, while Harrow was charging £38.40 per week rent and £7 per week Council Tax. Brent council was the most expensive and also charged rent and Council Tax on a temporary site at £68 for a double and £55 for a single pitch per week. Only six of nineteen local authorities provided play facilities, nineteen of the twenty-two local authorities surveyed provided water, electricity, flush toilets and refuse collections; sixteen provided amenity huts and ten offered a laundry service. Only Brent provided electricity on their temporary site. The report suggested a large number of Travellers are missing from official estimates because they rent private housing, while housed Travellers tend to be more nomadic than those living on permanent sites.

O'Dwyer's (1997) local work in London found that fifty percent of Irish Travellers surveyed on a Brent local authorities site felt that the site was unhealthy. Poor skin conditions and rashes were common among site residents after moving to the new LA site. Some residents blamed the landfill underneath the site, others pointed to problems with the site sewage system. Refuse collection on another site surveyed was said to be erratic causing problems with pests and other safety risks particularly for children playing near decomposing refuse. Some Irish Traveller women respondents suffered mental health problems while they also suffered from stress due to their children having no play areas and nothing to occupy their time.

Gaffney's (2001) local investigation into the needs of Travellers (mainly Irish Travellers) in the London Borough of Southwark gives a comprehensive overview of the social needs and lack of provision and support that characterise most LA responses to the needs of Irish Travellers and Travellers generally. Accommodation was the highest priority need felt by Travellers in Southwark. Overcrowding on three sites has become a serious fire hazard, sites lack basic amenities and repair, maintenance and cleansing was poor. Sickness and mortality rates are extremely high in the Traveller community with foetal and infant mortality rates running at between three and ten times the rates in the general population while accident rates were very high. Problems occurred when accessing emergency treatment and in getting diagnoses of disabilities at the primary care level. Disabled access to equipment and benefits appeared to be minimal while carers lacked adequate support. Racist abuse and a lack of vocational training for teenage Travellers contributed to extremely poor school completion rates. Child care responsibilities and difficulties in accessing nursery education were a serious barrier to the participation of female parents and childless young women in training and employment. Consequently unemployment and reliance on casual manual work was high. Young Travellers were particularly at risk of becoming embroiled in the criminal justice system. Travellers were ill-informed of their rights regarding many welfare and support services.

This research documents physical attacks on Traveller children, racist abuse and anti-Traveller graffiti, and the racially motivated rejection and abuse by settled neighbours of housed Traveller families. But Travellers were frequently unaware of the police and local authority support and sanctions available in such cases. Gaffney (2001: 6) concluded that:

> The degree of marginalisation and disempowerment shown in this research suggests that many of the services required by Travellers in Southwark need to earthed in community development. ... Community development will also enable the targeting of services and support a joined up approach to delivering them. However the community has little or no experience of voluntary sector practices or of project management. They therefore need resources dedicated to intensive support in the short and

> medium term. ... Their exclusion from regeneration funding to date indicates a considerable need to link Travellers into mainstream initiatives.

The above research reports concentrated on Irish Travellers generally and were located in London areas. A much larger body of research exists in England, Wales and Scotland that deals with other Traveller and Gypsy groups or collectively with all indigenous Travellers including Irish Travellers. This body of research usually reflects to some degree the social marginalisation and service disqualification suffered by Irish Travellers as outlined more specifically in the 'Room to Roam' report on which this paper is partly based, as does the considerable body of research on Irish Travellers in Ireland both north and south (see Clements and Morris, 2001). The next section of this paper looks at particularly vulnerable parts of the Irish Traveller population in England outlined by Power (2004).

Irish Travellers particularly at risk in Britain

Women: Irish Traveller women face discrimination as Travellers, as women, often as mothers and as Irish people in Britain. Hence, promotion of gender equality, both internally within their own strongly patriarchal community, and externally in the context of covert and overt discrimination and racism, is fundamental to their empowerment. However, most of the recent research work around women Travellers focuses on health issues. Specific health statistics for Irish Travellers are extremely scarce, partly due their relative ethnic invisibility. The health profile of all Travellers in Britain is significantly worse than other settled sections of British society. Stillbirth is seventeen times higher than the national average; infant mortality is twelve times higher than average rates – twice as many Traveller infants fall into the low birth weight category (Linthwaite, 1983). Traveller women's use of family planning; developmental screening; immunisation, and antenatal care services are very low in relation to other minority ethnic groups and British society in general (Pahl and Vale, 1986). Prenatal and antenatal eviction of women is thought to impact very adversely on the immediate and long-term state of their health and on that of their babies (Durward, 1990; Sadler, 1993).

Children: There is a lack of continuous primary and virtually no advanced secondary education and training among young Irish Travellers in Britain, while participation in further and higher education is extremely low. Traveller children are often bullied and verbally abused by other children and their defensive reaction can be to deny or disguise their identity or stick together, resist integration, fight back and trade racist insults and violence with opponents. The use by some Traveller children of violence and racist language results in high rates of temporary and permanent exclusion from schools, particularly those that do not value or recognise their ethnicity, while settled children who use 'unrecognised' racist terms like ‘gippo’, 'pikey', ‘knacker’ etc. often go without reprimand by authority figures. These instances indicate either and unwillingness to recognise or a lack of awareness of Travellers ethnicity and culture by many school administrators, teachers and pupils (Power, 2004). A high incidence of congenital malformation occurs in traveller children – common assumptions link this to high rates of inter-marriage, but little research has been carried out into the possible socio-economic and environmental factors that may have an influence (Linthwaite, 1983; see Power, 2004, chapter 3).

Young men: Many traditional rurally based Traveller economies have disappeared and rapid urbanisation has eaten away at the social fabric of the extended Traveller family as unsuitable ‘nuclear’ family housing physically divides up the extended family and can lead to a breakdown of internal group discipline. Due to tight restrictions and discipline within families young men (and occasionally women) break away from Traveller constraints and conventions in urban situations. High rates of marriage break-up and alcohol abuse due to cultural breakdown result in low self-esteem and can lead to criminalisation and violence within and between Traveller families. Young male Travellers become ‘youth at risk’ as they rarely attend secondary school and spend much time on the streets where some come into contact with drugs, gangs, petty theft, car crime and attract the attention of the police as ‘street people’. This often results in imprisonment in Young Offender Institutions sometimes progressing eventually to adult prisons due to presumptions of a ‘transient lifestyle’ by criminal justice officials. Lack of a ‘fixed abode’, poor culturally insensitive or overtly prejudiced pre-sentence reports by probation officers, and Travellers’ own ignorance of the judicial system result in a high proportion of custodial sentences (Devereaux, 1999; Power, 2003b and 2004). A combination of machismo from some staff, Irish Travellers and other inmates, and anti-nomadic prejudices from prison officers and other inmates often results in fights and disobedience in prison and consequent high rates of disciplinary measures and sanctions against Irish Travellers (Power, 2004, chapter 5).

The continued urbanisation and criminalisation of Irish Travellers has impacted severely on the lives of young males in particular. Social alienation, illiteracy, and low self-esteem, make it difficult for young males to access education, training, and leisure facilities in urban areas. They spend much of their time in overcrowded accommodation or on the streets. High levels of boredom lead to tension and friction in both these arenas. Young males are extremely prone to discrimination as Travellers, Irish people, and as young men. A report by the London Gypsy and Traveller Unit (Witt, 1995: 1), a voluntary Traveller support group in London, asserted that young male Travellers:

> …meet prejudice and intimidation from many quarters but particularly from the police. They are a highly stigmatised group that are sometimes turned away from leisure facilities and shops because they are perceived to be thieves, dirty and unruly.

The relationship between the Irish Traveller community, the settled population in general, and the police in particular is one that is often characterised by hostility, prejudice, discrimination, ignorance, and conflict in Britain (Power, 2004; Crawley, 2004).

British state policies and the assimilation of Irish Travellers

Repressive anti-Traveller laws were only repealed in England and Ireland during the eighteenth century. In the nineteenth century Irish Travellers (particularly after the Great Famine in the late 1840s) emigrated from Ireland to many countries including Britain, the United States and Australia though Irish Travellers have traditionally plied regular migratory routes between Britain and Ireland over the centuries (McDonagh and McVeigh, 1996; Irish Travellers cited in Power, 2004). In the twentieth century economic changes made many Traveller trades obsolete while urbanisation and planning laws closed numerous traditional stopping places to Travellers throughout Britain. Economically successful Irish Travellers bought sites to live on, but this option was severely curtailed by the 1960 Caravan Sites (Control of Development) Act in Britain.

The majority of Irish (and other) Travellers struggled to adapt their economies successfully in the new post-War climate of rapid economic change and expanded government sponsored social planning and development in Britain that took either negative or no account whatsoever of indigenous nomadic needs. Consequently, nomadically based groups including Irish Travellers, were plunged into a crisis for survival. The

Irish Traveller population in Britain also increased during this period as emigration from Ireland reached its zenith during the late 1950s. Rapid population growth since then has led to an extremely youthful demographic profile amongst Irish Travellers, while urbanisation has placed huge strains on the extended family unit sometimes resulting in the breakdown of traditional social structures and constraints with nothing tangible to replace them – at best the relevant authorities conveniently assumed a painless assimilation process had occurred. However, high mobility rates, coupled with the dearth of stopping places and the criminalisation of nomadism have brought many Irish Travellers increasingly into reluctant conflict with the police and the criminal justice system (Power, 2004).

Attempts by the British state and its institutions to assimilate Travellers into the settled population during the twentieth century have taken various forms. This has included overt and covert coercive legislation that has made it increasingly difficult for Travellers to use traditional stopping places, or access authorised sites and services. Local authorities have increasingly abandoned their responsibility to provide health, education and welfare services to excluded transitory Travellers since the 1968 Caravan Sites Act (see relevant chapters in Power, 2004). Some legislation has also unintentionally impacted negatively on the lives of Travellers including the Public Health Act 1936, the Town and Country Planning Act 1947, the Highways Act 1959, and the Caravan Sites and Control of Development Act 1960 (Hawes and Perez, 1995: 56). The Environmental Protection Act 1990 (EPA) criminalised unregistered scrap metal collection, and consequently further undermined one of the major economic occupations of Irish Travellers (O'Nions, 1995: 10, accessed on 21/6/2002 http://webjcli.ncl.ac.uk/articles3/onions3.html). The dearth of adequate temporary and permanent site provision is a primary determinant of the extremely poor life-chances of Irish Travellers including shockingly low health and education indicators, escalating criminal justice issues, and a deteriorating nomadic economy. Indigenous forms of nomadism have gradually been outlawed in Britain over the course of the twentieth century.

The Caravan Sites Act (CSA 1968) and the Criminal Justice and Public Order Act 1994 (CJPOA) are the most recent legislative manifestations of the state's negative attitude to nomadism. The CSA (1968) made it a statutory obligation for local authorities to provide permanent camp sites for all Travellers 'residing or resorting' in their particular area (while also exempting some local authorities from the need to provide sites). Many of these new local authority authorised sites built since the 1970s were poorly or dangerously situated close to polluted motorways, flyovers, industrial estates, hazardous technology (for example electricity pylons), and in poorly serviced and inaccessible areas (CIEH, 1995; Sibley, 1995). Hawes and Perez (1995: 29) outlined the salient defects inherent in the CSA (1968):

> First, the initial assessment of numbers and needs was, with hindsight, lamentably underestimated; second the Act is concerned with site provision only, disregarding issues of social provision; and third it failed to set a time-scale within which the mandatory duty to provide sites was to be completed.

When local authority site provision reached 'Designated' status under the CSA (1968) their territorial areas were virtually closed to transitory nomadic Travellers who often happened to be 'new' Irish Travellers.[9] The Act was deeply flawed on its demographic premise, the nature of the accommodation prescribed, the absence of a holistic approach, and at an operational level. With an unsatisfactory level of enforcement the CSA (1968) was interpreted as loosely or inadequately as was deemed appropriate according to the preoccupations and prejudices of particular local authorities. The predictable result was a continuous and escalating shortage of site provision coupled with accelerated commercial encroachment on traditional Traveller stopping places, and a continuation of unauthorised Traveller camping coupled with increasing friction with sections of settled society as the Traveller (and settled) population both increased and concentrated around urban centres. The CJPOA's (1994) pernicious impact on indigenous nomadic people resulted in the criminalisation of Traveller's nomadic lifestyle and the further undermining of their various economies (Turner, 2000). The CJPOA (1994) abrogated the (already weak) statutory obligation of local authorities to provide some local site facilities for Travellers and moved the British state's deeply flawed assimilationist Traveller policies into a coercive trajectory that now virtually outlaws most forms of indigenous nomadism (Power, 2004).

---

[9] Settled Irish immigrants to Britain after the War were often regarded with deep suspicion, as a result Irish Travellers attracted similar opprobrium coupled with settled antipathy to indigenous nomadism, reinforced with imported prejudice from many in the new settled Irish communities in Britain.

State mediation of anti-Irish Traveller racism

Britain's indigenous nomads (usually semi-nomadic even when supposedly settled) include the largest group Gypsies (also known as Romanies - with Welsh, English and Scottish subgroups), Irish Travellers, Scottish Travellers, fairground, circus and show people, and 'New Age' Travellers. In the post Cold War period significant numbers of Roma (European Gypsies) have migrated from the former Soviet bloc to Britain due to systematic discrimination and persecution in the new liberal democratic states of central and eastern Europe. Confusion reigns at all levels of settled society as to 'who or what are Travellers?' Colloquially, derogatory terms like 'Gypsy', 'Tinker', 'Knacker', and 'Pikey' are often used incorrectly (both ignorantly and/or pejoratively) to describe Irish Travellers at all levels of settled society. British state discourses and policies in relation to indigenous nomadic ethnic groups like Irish Travellers and Gypsies are contradictory and sometimes betray a more sinister trajectory and logic centred around notions of 'acceptable' and 'degenerate' ethnic group identities – romanticised 'elect' Romanies and 'reprobate' Irish Travellers (Ibid).

This confusion as to who Irish Travellers are is exacerbated by hierarchies of authenticity generated historically by economic competition, ethnic and national rivalries, ignorance, and prejudice between indigenous Traveller groups – all of which are exploited by some politicians at all levels of government in Britain. Historically British Travellers and Gypsies who claim Romany origins are accorded (and sometimes claim) a privileged nomadic and essentially romanticised ethnic 'authenticity' denied their Irish nomadic counterparts (and other indigenous Travellers) in official discourses ranging from the academic to political arenas. Irish Travellers eschew a Romany connection, though some opportunistically adopt common English Traveller sobriquets when accessing services in Britain – proof itself of this pervasive hierarchy of authenticity (Irish Travellers cited in Power, 2004; Okely, 1983: 15). This hierarchy of authenticity and its concomitant prejudices are replicated throughout society's various engagements with Irish Travellers.

Levitas (1998: 178) outlines the contradictions, multiple interpretations and conflicting policy options that the adoption of a social exclusion agenda affords governments allowing them to create a hierarchy of legitimacy of need that then influences any policies and support offered:

> Social exclusion is a powerful concept, not because of its analytical clarity which is conspicuously lacking, but because of its flexibility. At an individual level, it mobilizes personal fears of being excluded or left out,

> which reach back into childhood as well as having immediate reference. At a political level, it has a broad appeal, both to those who value increased participation and those who seek greater social control. Crucially, social exclusion facilitates a shift between the different discourses in which it is embedded, so that the contested meaning of social exclusion now lies at the heart of political debate. The boundaries of this debate are set both by the real configuration of political forces and by the language in which it takes place. That language reflects and reproduces underlying ideas about what society is and how it works.

In this context New Labour's variable concept of 'social exclusion' can tend to encourage opportunist political 'spin' at the expense of substantial policy and practice change in Britain. The janus faced nature of New Labour's ideological and policy response in the context of 'social exclusion' concepts to inequality, poverty, discrimination and racism, and to Irish Travellers' structural position and needs in particular is exemplified by juxtaposing some of the contradictory views of the powerful politicians including Labour ministers quoted below. In an interview with BBC Radio West Midland in July 1999 then Home Secretary Straw (cited in Guardian, 20 August 1999) made a wide ranging attack on 'non-Romany' Travellers:

> There are relatively few real Romany gypsies left, who seem to mind their own business and don't cause trouble to other people, and then there are a lot more people who masquerade as travellers or gypsies, who trade on the sentiment of people, but who seem to think because they label themselves as travellers that therefore they've got a licence to commit crimes and act in an unlawful way that other people don't have.

Rachel Morris (cited in ibid), from Cardiff University's Law School, commented on Straw's motives: *'He is a very canny politician ... He knows that he will be able to get away with this.'* Straw's outburst also illustrates how the 'hierarchy of authenticity' is articulated at the highest levels of national power to de-legitimise certain Traveller groups, while also undermining pan Traveller solidarity by reproducing the concept of 'elect' and 'reprobate' Traveller populations. In effect, New Labour's ideology of social inclusion has reinvented the Victorian concepts of the 'deserving' and 'undeserving' poor, but applied in this particular context to two recognised ethnic groups both with origins rooted in indigenous nomadism (Levitas, 1998; Power, 2004).

In a damage limitation exercise following Straw's outburst, a Home Office spokeswoman (cited in Guardian, 20 August 1999) maintained that:

> His comments have been misconstrued. He was referring to people who are engaging in criminal activities and anti-social behaviour, not to the traveller population as a whole. The government has a policy of toleration towards those who choose this lifestyle but that does not extend to crime or disruption of the local community.

This Guardian report ends in a curious fashion by switching incongruously to an unrelated controversy about an Irish Traveller wedding and the consequent temporary closure of pubs in two small towns in Ireland. The newspaper ends the Straw article by quoting a spokesman for the Vintners' Federation of Ireland: *'These people [Irish Travellers] seem to think they have a right to fight in pubs and wonder why they are not being served.*' Even when social democratic politics meets left liberal journalism, the limits of liberal pluralism are exposed in relation to Irish Travellers in Britain.

As the above excerpts from ministers' discourse illustrates the Home Office tolerates rather than celebrates ethnic groups whose cultures include nomadic characteristics. A correspondent to the Guardian Letters page (Imrie, 21 August 1999) placed Straw's comments in the recent historical context and continuum of the disqualification, racism, de-humanisation and genocide inflicted on European nomads in the middle twentieth century:

> Jack Straw's outburst about "so-called travellers" has a long pedigree, as I discovered reading the new English translation of Otto Rosenberg's first-hand account of the persecution of Gypsies by the Nazis, A Gypsy in Auschwitz, published at the end of this month. In 1936, Dr Robert Ritter, director of the "Racial-hygienic and Genetic Research Office" claimed that many so-called Gypsies were "half-breeds" and should be distinguished from "racially pure Gypsies". It was not clear who belonged to the latter category, but the former, having "inferior genes", were "highly unstable, unprincipled, unpredictable... lethargic or restless and irritable... work-shy and asocial". Surprisingly, perhaps, he didn't actually claim that they defecated in doorways. The distinction between real and fake Gypsies made little difference in the concentration camps where many thousands of them were murdered indiscriminately.

Some months after Straw's outburst home office minister Boateng (cited in Monbiot, Guardian 4 November 1999) informed parliament that '*the home secretary's remarks did strike a chord throughout the country [accordingly he was considering strategies to] strengthen the hand of both local authorities and police*'. Monbiot (ibid) suggests that this was yet another attempt by New Labour to not only maintain but enhance its 'law and order' image with 'Middle England' (essentially the white 'indigenous' conservative middle-classes) by stigmatising even further extremely marginalised Traveller groups who have historically been subjected to forced assimilation and even genocide.

Turner (Guardian, 24 August 1999) outlines how the language used by some politicians and media coverage not only negativises Travellers but through the use of aggressive heightened language constructs the meeting of sedentary and nomadic lifestyles in a one dimensional negative manner:

> The terminology used in reporting stories about gypsies and travellers [sic] is often conflictual and derogatory to both travellers and their supporters. When gypsies arrive anywhere in numbers, they "invade". A proposal for a new caravan site sees local residents "battle" to stop it. If they succeed, they score a "victory". It is the language used about enemies.

By using such warlike and emotive language a state of endemic conflict is seen to exist between Travellers and sedentary society and also state and local institutions embedded in sedentarism. Official discourse at all levels has rooted in it this presumption that Travellers have a criminogenic disposition. When a British Home Secretary feels free to pronounce on public radio that 'certain' kinds of Traveller are illegitimate (and receives no reprimand or sanction from parliament, party or prime minister), it is hardly surprising that ordinary policemen feel it is legitimate to stereotype Travellers as a criminogenic group and police them accordingly (Power, 2004, see chapter 5). Whatever happened to New Labour's 'social inclusion' mantra in this case?

Travellers generally, and Irish Travellers in particular, though a relatively small minority in numerical terms in Britain, seem to have become one of the main indigenous recipients[10] of the most negative and racist discourses that reaches from ministerial heights, through many institutional levels and often manifests as overt discrimination operationally (police cited in Power, 2004). Conservative MP Andrew Mackay (cited in The Guardian 15 January 2002), a former shadow Northern Ireland secretary, moved the 'debate' towards a dangerous dehumanisation of Travellers when he railed against Traveller "invasions" of his constituency insisting that these Travellers did not deserve the same human rights as his constituents and he asked whether the police were *'doing a proper job'* in a Westminster Hall debate:

> Ordinary, innocent people, hard-working, normal, straight-forward people, who live around Bracknell want to get on with their lives in peace. But they want to have protection under the law when invaded by the scum

[10] Refugees and asylum seekers are open to similar abuse.

> [Travellers]. And they are scum, I use that word advisedly, because people who do what these people have done do not deserve the same human rights as my own decent constituents going about their everyday lives.

In reply Junior Home Office minister Angela Eagle (cited in ibid) condemned Mackay's 'scum' reference and continued:

> We do not recognise that the minority status of travellers and gypsies [sic] should allow them to indulge in crime or anti-social behaviour or excuse that behaviour if they do. The law should be enforced equally towards them.

Eagle's rejection of the 'scum' label for Travellers generally is followed by an insidious discourse linking criminal and anti-social behaviour to Traveller and Gypsy ethnicities. Again the inference is that these are criminogenic groups. Eagle infers that the all 'citizens' should be equal before the law. But the law is not equal in respect to Irish Travellers and other ethnic groups with a nomadic disposition. For example, the CJPOA (1994) is not enforced equally with respect to them – it criminalises nomadism which is a central tenet of all their cultures, whether actual or aspirational. What other ethnic minorities could a government minister 'committed' to social inclusion label in such negative terms? What does this say about the limitations of New Labour's social inclusion policies and Britain's presumed ethnic and cultural pluralism (Power, 2004; Levitas, 1998)?

Conclusion

This paper explains how Irish Travellers came to be a recognised indigenous ethnic group in Britain through historical migration routes and latterly as more permanent immigrants. It also outlines what delineates Irish Travellers culturally uniqueness in Britain in relation to indigenous sedentary society, settled Irish immigrant communities and other groups with an indigenous disposition, while also attempting a demographic overview of the population. This leads to a discussion of why the 'Room to Roam: England's Irish Travellers' report on which this paper is based was commissioned. This research was concerned with the social exclusion and disadvantage in terms of accommodation, health, education, welfare, and criminal justice of a severely disadvantaged and marginalised minority in British society. This research explored ways of developing and providing more reflective and culturally sensitive institutional policy and practice in key areas of service delivery, including strategies that promote Irish Traveller empowerment and influence in the design and delivery of key services see Power, 2004). The small amount of previous specific research on Irish Travellers in Britain is reviewed and critiqued and the social situation of the most 'at risk' sections of the Irish Traveller population in Britain, women, children and young men, are discussed. This burdens these sub-groups with distinct disadvantages in accessing health care, educational support, and social welfare. In some cases the breakdown of their cultural supports, particularly the extended family, linked to the criminalisation of nomadism and the undermining of Irish Traveller economies has led to them being over-represented in the Criminal Justice System.

As demonstrated by comments from politicians in the latter section above Irish Travellers are discriminated against and excluded both institutionally and socially, and as such are one of the most marginal and misunderstood ethnic groups in British society. The predominant (but confused and contradictory) trajectory of recent government policy has largely been towards assimilation into settled society. The actual history of Irish Travellers in Britain, who are commonly linked in both many settled British and Irish psyches to extremely negative stereotypes, combined with their 'whiteness', means that they fit uneasily or are altogether ignored consistently in central government policy formulations and initiatives centred on issues of ethnicity and diversity. Moreover, even since the official recognition of Irish Travellers as an ethnic group in Britain in 2000 and the advent of the Race Relations Amendment Act (RRAA 2001) required all public authorities to include all recognised ethnic groups in race equality schemes and official discourse. In relation to the RRAA the Discrimination Law Association (2002: 219) assert that:

> 'Local pressure will be needed to ensure that the needs of Romany Gypsies and Irish Travellers are addressed. Authorities may need reminding that both are racial groups under the Act.'

Without this recognition and ethnic monitoring for Irish Travellers both respect for their ethnicity and culturally appropriate service provision will continue to be piecemeal and patchy at best (Power, 2004; Parekh, 2000: 139). The failure to include 'Irish Traveller' as an ethnic category in the Census (2001) is indicative (and partly a cause) of the institutional blindness that Irish Travellers endure as an ethnic group and of their outcast, even outlaw status in élite, state, institutional and colloquial discourses and practices in Britain. Whilst affording Irish Travellers ethnic recognition through judicial decisions, the British state simultaneously criminalises them through anti-nomadic legislation and persecutes them by the negative discourse and discriminatory actions (or inactions) of a range of state actors and public institutions.

References

Acton, T (1994) 'Categorising Irish Travellers', in M. McCann et al, *Irish Travellers: Culture and Ethnicity*. 36-53, Belfast: Queen's University.

Adams, B; Okely, J; Morgan, D and Smith, D (1975) *Gypsies and Government Policies in England. A Study of the Travellers' Way of Life in Relation to the Policies and Practices of Central and Local Government*, London: Heinemann.

Binchy, A (1994) 'Travellers' Language: a sociolinguistic perspective', in McCann, M., *Irish Travellers Culture & Ethnicity*. Belfast: The Institute Of Irish Studies, QUB.

Cauley, W. (2005) *Canting with Cauley, A Glossary of Travellers' Cant/Gammon,* Transcribed and edited by M. Ó hAodha, Dublin: A. & A. Farmar.

CIEH (1995) *Travellers and Gypsies - an alternative strategy*. London: Chartered Institute of Environmental Health.

Clarke, N (1998) 'Living on the edge of your town: A study of Traveller provisions and conditions in outer London and the home counties.' In Egan, K and Wake, M (eds.) *'BRIDGE',* London: Spider pub's.

Crawley, H (2004) *Moving Forward – the provision of accommodation for Travellers and Gypsies*. London: IPPR.

Devereaux, D (October 1999) *Enforced Invisibility: The Irish Experience of the Criminal Justice System.* MA thesis submitted to the School of Law, Manchester Metropolitan University: unpublished thesis.

Discrimination Law Association (2002) *Briefings*, February, Volume 15.

Durward, L (ed.) (1990) *Traveller Mothers & Babies: Who cares for their health?* London: Maternity Alliance.

Fraser, A (1992) *Gypsies*. Oxford, Blackwell Publishers.

Gmelch, G and Gmelch, S (1985) 'The Cross-Channel Migration of Irish Travellers', *The Economic and Social Review*, Vol 16., No. 4., July, pp287-296.

Guardian newspaper – various news reports and letters to the editor (see text), London.

Hawes, D and Perez, B (1995) *The Gypsy and the state*. Bristol: SAUS Publications.

Gaffney, M (2001) *Needs analysis of the Travelling Community in Southwark*. London: Southwark Social Inclusion Unit.

Hickman, M and Walter, B (1997) *Discrimination and the Irish Community in Britain*. London: CRE.

Hyman, M (1989) *Sites for Travellers: a study in five London Boroughs* London: London Race and Housing Unit.

L'Esteve, D (1986) *Travellers and Welfare Benefits*. London: London Strategic Policy Unit.

Levitas, R (1998) *The Inclusive Society? Social Exclusion and New Labour*. New York: Palgrave.

Linthwaite, P (1983) *Health and Health Care in Traveller mothers and children*. London: Save the Children Fund.

LIWC (1995) *Rights for Travellers - A LIWC Survey of Local Authority provision for Travellers in London*. London: London Irish Women's Centre.

McCann, M; ỌSíocháin, S; and Ruane, J (Eds.) (1994) *Irish Travellers: Culture and Ethnicity*. Belfast: The Institute Of Irish Studies, The Queen's University of Belfast.

McCarthy, D and McCarthy, P (1998) *Market Economies: trading in the Traveller economy*, Dublin: Pavee Point Publications.

McDonagh, M and McVeigh, R (1996) *Minceir Neeja in the Thome Munkra (Irish Travellers in the United States)*, Belfast: Belfast Travellers Education and Development Group.

Monbiot, G (4 November 1999) Appealing to the worst in all of us, in the *GUARDIAN*.

Morris, R and Clements, L (2001) *Disability Social Care, Health and Travelling People*, Cardiff: University Press.

Niner, P (2002) *The Provision and Condition of Local Authority Gypsy/Traveller Sites in England. A Report to Office of the Deputy Prime Minister on a Programme of Research*. Centre for Urban and Regional Studies, University of Birmingham and Office of the Deputy Prime Minister, London.

Ó Riain, G (1997) (ed.) *Travellers: Nomads of Ireland*, Dublin: Pavee Point Publications.

O'Dwyer, M (1997) *Irish Travellers Health Access Project Draft Report.* London: BIAS Irish Travellers Project and NHS Ethnic Health Unit.

O'Nions, H (1995) 'The Marginalisation of Gypsies'. Website accessed on 21/6/2002, http://webjcli.ncl.ac.uk/articles3/onions3.html

ODPM (2001) 'Gypsy/Traveller Counts' www.housing.odpm.gov.uk/information/index14.htm, Office of the Deputy Prime Minister.

Okely, J (1983) *The Traveller-Gypsies*. Cambridge: University Press.

—. (1994) 'An anthropological perspective on Irish Travellers', M. McCann et al (eds.) *Irish Travellers: culture and ethnicity*, Belfast: The Institute of Irish Studies, QUB.

Pahl, J and Vaile, M (1986) Health and Health Care Among Travellers. Kent: University of Kent, Health Research Unit.

Parekh Report (2000) *The Future of Multi-Ethnic Britain. Report of the Runnymede Trust Commission on the future of multi-ethnic Britain.* London: Profile Books Ltd.

Power, C (2003a) 'TELLING IT LIKE IT IS'? Power, prejudice, politics and people in the qualitative process', in S. Tombs, and D. Whyte (eds.) *Unmasking the Crimes of the Powerful*, London and New York: Peter Lang.

—. (2003b) Irish Travellers: Ethnicity, Racism, and PSRs, *Probation Journal,* Vol. 50, September Issue 03.

—. (2004) *Room to Roam: England's Irish Travellers - a Report of Research*, London: Community Fund.

Rutter, P (1997) *From the Margins to the Mainstream: An Effective Management Strategy to increase access by the Haringay Traveller community to Council Services.* Haringay Council: Work based Project.

Sadler, C (1993) 'Out in the Cold', in *Nursing Times*, 89, (16): 16-17.

Sibley, D (1995) *Geographies of exclusion - Societies and Difference in the West.* London: Routledge

Turner, R (2000) 'Gypsies and politics in Britain'. *The Political Quarterly*. 71:68-77.

—. "Jack Straw is alone in his certainty about who a gypsy is", The *Guardian;* 24 August 1999)

Witt, S (Sept. 1995) *LGTU Report on their Groupwork with Traveller boys (Fianna Buachailli).* London: London Gypsy and Traveller Unit.

CHAPTER NINE

AN ILLUSTRATIVE STUDY OF SOCIAL SERVICES DEPARTMENTS (SSDs), MIGRATION, HOMELESSNESS AND YOUNG IRISH TRAVELLER CHILDREN IN ENGLAND

COLM POWER

Introduction

This paper is based on primary qualitative research conducted as part of 'Room to Roam: England's Irish Travellers' (Power, 2004), a Community Fund resourced research project.[1] This three-year project involved focus groups and many in-depth interviews with Irish Travellers and relevant service providers (health, education, social services, politicians, voluntary services, police and the criminal justice system) that were carried out for the most part in London and Manchester. The names of respondents are disguised throughout this chapter in order to protect their anonymity. The primary data used in this chapter draws principally on a small number of in-depth interviews with Irish Travellers in England, and also an in-depth interview conducted with the Travelling People's Team (TPT) at Haringey Social Services Department (SSD) in London, while the central narrative of the piece is taken from and in-depth interview with Mr. Tarrington and Outreach Worker for the voluntary charity group Safe City based in central Manchester. The 'Room to Roam' research aimed generally to assess the impact of social marginalisation,

---

[1] The research project was run by a consortium that included Action Group for Irish Youth, BIAS Irish Travellers Project, and University of Surrey - St. Mary's College.

ethnic disqualification, and criminalisation on Travellers' health and life-chances and to suggest positive policy and practice developments.

This chapter links academic and policy critiques of service provision and delivery by SSDs to Irish Travellers and their children in particular, and other nomadically based groups generally, and then illustrates these critiques through the narrative experiences and insights of the respondents referred to above. The social context for this chapter lies in the extreme stresses and strains on the culture, nomadism and social organisation of many contemporary Irish Traveller families that is at least to some significant degree a consequence of the assimilation and criminalisation policies pursued in relation to indigenous nomadic groups by successive British governments over the past 40 years (see Power, 2003). The chapter then examines some very specific issues relating to institutional ignorance and neglect experienced by homeless Irish Traveller children in the delivery of statutory social services in some areas of England. It also highlights good practice in SSDs and related initiatives where these could be identified.

Vulnerable children, Travellers and SSDs: the context

Cemlyn (1998) comes to the interesting conclusion that Travellers may have much less need of orthodox SSDs than settled people because extended family supports and supportive relationships between families can provide a comprehensive community safety net. What Travellers do need, however, is culturally sensitive support and resources to bolster and maintain the extended family itself particularly when it comes under the severe stresses of urbanization and settlement in inappropriate forms of accommodation, whether it be on overcrowded purpose-built sites or in family housing designed around the nuclear family model. But Criminal Justice and Public Order Act (CJPOA, 1994) related evictions and prosecutions, while undermining the extended family, have increased SSDs involvement with Travellers due to welfare issues, though few Social Service Departments have developed any specific policies or good practice in this area – the Travelling People's Team (TPT) that operates in Haringey Borough in London being a notable exception (TPT interview cited in Power, 2004). But where Travellers come into contact with SSDs at their most vulnerable is when living in inappropriate forms of accommodation such as conventional housing or when actually homeless. Lack of awareness about Traveller culture in SSDs can lead to serious misunderstandings that exacerbate problems especially in crisis intervention situations. Families may be de-stabilised and pathologised by insensitive professional treatment that undermines the very family supports that SSDs purport to support. This partially explains why Travellers often react by moving-on due to the intrusive involvement of SSDs (Cemlyn, 1998; Tarrington and Irish Travellers cited in Power, 2004). Government and media negative misrepresentations of Travellers also permeate into SSDs and contribute to prejudicial, insensitive and poor quality engagement and service delivery (Traveller Law Research Unit, 1997).

The geographical breakdown of statutory SSDs into regional and council boundaries militates against support for nomadism generally and Irish Travellers specifically in England. There is unwillingness by SSDs to acknowledge responsibility for transient populations due to bureaucratisation, self-imposed territorial limitations, under-staffing, poor morale and lack of awareness about Travellers. But also, SSDs will usually only engage with 'at risk' people on their own set institutional terms and criteria. Manchester SSD lacks co-ordination and co-operation with Traveller Education Services (TESs), other relevant statutory bodies and specialist Traveller voluntary sector support organisations (Tarrington and Stryder interviews, Manchester Interdisciplinary Forum discussion, all cited in Power, 2004). Recent research carried out into the 'management' of illegal Traveller encampments in Britain noted that SSDs were the least likely of all local statutory welfare agencies to visit these sites (Cowan and Lomax, 2003: 307). This situation is somewhat peculiar given SSDs remit to support and protect families and children. This may be partly explained by the deep antipathy to Social Workers felt by many Irish Travellers (and other ethnic Traveller groups also) – they are often regarded as 'child stealers' (ibid). Though Garrett's (2003) research into the treatment of Irish families and children was actively supported by the Association of Social Services Directors only forty-four percent of SSDs contacted in England and Wales responded to his questionnaire. Of these, forty-two percent claimed to have specific services for Travellers while the other SSD respondents did not provide any culturally sensitive support for Travellers.

Lord Laming, a former director of Social Services and chair of the inquiry into Victoria Climbié's death challenged the government to act on his report (2003).[2] Lord Laming's (cited in Perkins, Guardian: 28 March 2003) evidence to the Parliamentary cross-party health committee warned

[2] Eight-year-old Victoria arrived in Britain from Ivory Coast and died in February 2000 after ten months of severe abuse inflicted by her guardian despite being referred to locally to social services, a hospital and the Metropolitan police.

there would be more 'children at risk' deaths unless crucial changes were made to SSD higher management practice:

> I hope that before any of these people in key positions think of going on summer holidays they will have satisfied themselves that those recommendations that should be in operation are in operation. … It's no use parliament producing good legislation ... if parliament can't be sure that legislation is implemented at the front door. That is the acid test - it's about whether children and families get a good service. I believe there is a yawning gap between the aspirations and expectations of parliament and what is delivered at the front door. … But, if those principles … are not addressed then I think there is every likelihood that there will be more inquiries like mine. … The safety of children and the well-being of staff should not depend on protecting incompetence in management. … Why are these well-intentioned people [frontline social workers] put in the situation where they felt defeated by the task in front of them? … We need to rediscover the basic principle that the public services are there to serve the public and not just some people who can get through eligibility criteria.

Dennis (cited in Power, 2004: 49), a young male Irish Traveller (this is not his real name) incarcerated in Feltham Young Offenders Institution, came from a family living in settled accommodation in London. Below he explains how his background, family relationship breakdowns and deaths in his extended family contributed to his mental health problems:

> It's bad in a way 'cos [father] just got up and left like that … I don't think that would have happened in Ireland. … [W]hen I come to England that's when it all went wrong, they separated straight away. … [Big cities] don't help families. I think for a businessman this is the best place 'cos they make plenty money, but it's no place for a family … there is too much pressure... I have lost a few uncles through … sudden deaths and one of them got murdered. I was under a bit of stress, that's what done my head in. … But when you lose that person in your life it's like losing your whole life because … I was always brought up with my uncles ... now that they are gone I just feel a bit lost...

The break-up of his immediate family and the loss of important male role models in his extended family destroyed much of Dennis' social support network and his sense of self and being. Dennis (Ibid) describes his first encounters with SSDs in his early teens as he increasingly rejected secondary school and got involved in petty crime:

> I was going to school … I just met these couple of boys who said: "Let's go out and do this", and I thought ... "What's the harm?" ... I was doing cars and silly little things like … shoplifting. I was only fourteen and I was thinking: "I can do what I like, I don't care." I would be walking around in brand new clothes and … my mother thinking: 'Where is he getting all the money from?" ... My family was getting too worried about me... Then the school started getting on top of me as well 'cos I would be taking two weeks off school. ... They would ring up me mother and she would say: "Dennis wasn't sick." We ended up in the head's office telling the school governors … the truth because my mother was next to me and I couldn't lie… That's when the social workers … come down on me, that's when I start getting into it deep.

Dennis' (Ibid) loyalty and respect is given almost exclusively to his extended family – the extended family form the limits in his social world. By taking a coercive and intrusive approach the intervention by SSDs exacerbated an already deteriorating situation. Instead of sensitively supporting the family's existing internal control structures to encourage a change of attitude SSDs dictated the agenda forcing the whole family to flee as Dennis relates:

> They were going to pick me up every morning to bring me to school and then get me after school to drop me back home, but that never happened 'cos my mother moved away. ... I didn't carry on [stealing]. I started working. Would you ever use SSDs? 'No, I have thought it over [about] Social Services. [T]hey want to know too much about you, and then … they take over your life. [B]ut the charities … they don't want to know about your life, it's direct help instead of like trying to get information... [SSDs] start throwing in the police ... you will be out on the streets. I don't trust [SSDs]... [voluntary services] are the best people to have.

Dennis makes a direct link between his disaffection with school, the consequent involvement of SSDs followed by increasing police interest in his activities that eventually resulted in a spell in youth prison. His experience confirms many Irish Travellers preconceptions and suspicions about the motives and remit of SSDs, while he feels that dedicated voluntary sector support is usually more culturally sensitive and less intrusive. The most worrying aspect of Dennis' experience is that instead of a sensitive intervention to support a family who were proactively willing to engage with his school to curb Dennis' truancy and petty crime: the school, SSDs and then the police tried to impose a coercive regime without sensitivity to his culture or situation and without effective consultation. The family relocated to avoid insensitive official interference, Dennis didn't get the support he needed, and eventually ended up serving a custodial sentence. This antipathy to SSDs runs deep with many Irish Travellers as a Traveller male of thirty (cited in Power, 2004: 49) relates: '*My two sisters was taken into foster care ... for maybe eight years and ... when my mother died the Social Services wouldn't let them attend the funeral...*'

SSDs and homeless Irish Traveller children

The illustrative study that follows underlines Lord Laming's point about the remit of SSDs being to serve the public and not the 'eligibility criteria' invented by its own bureaucracy. Unfortunately, this narrative is also another indictment of SSDs policy and practice in England. This study is also fascinating from the perspective of doing outreach and support work with so-called 'hard-to-reach' groups as it documents how a voluntary support organisation specialising in homelessness work struggled to find a working relationship with Irish Traveller children, a group the had never previously encountered in their work. Mr. Tarrington (cited in Power, 2004: 50) is a voluntary sector project worker with "Safe City" project (both the project and the worker's names are disguised here) that works with homeless people and he describes his project in Manchester:

> The … project has been in Manchester for about thirteen years [and] specifically works with young people who have run away [and] … are at risk on the streets… The young Travellers that we worked with we picked up through street work. … I first met their family working for another organisation in Manchester doing street work and I used to know a lot of the older brothers and cousins. I did resettlement work … attempting to find them accommodation, which wasn't largely successful. I was working with older [Traveller] lads. We now work with seventeen and under … so through street-work we met them at twelve, thirteen and worked with them up to about fourteen, fifteen. As a group of young men they were very suspicious, very hard to engage, [and] as very chaotic. We would see them out and about in town and give them resources … for a number of months and then they began to come to the building here.

Tarrington confirms here the deep suspicion that many Travellers have of outreach workers and the need initially to concentrate on the building up of trust over a period of time. The boys' nomadic circuit between Manchester, Birmingham and Dublin (and possibly other cities and locations in Ireland and Britain) made it difficult for Safe City to maintain the continuity of relationships with a range of Irish Traveller children who surfaced in Manchester periodically. As Tarrington (Ibid) relates:

> There were others [Travellers] who popped in and out, but they [twelve-year-old Irish Travellers twins] were the two constants - their pattern of movement was between LMC, Manchester and Dublin. [T]hey were born in Manchester and they had family who were relatively settled for a time. They were known to Social Services [in Manchester and Birmingham] through their previous addresses in Manchester, and then through bereavements etc., the family split up. They had surviving relatives, an older sibling in Birmingham … would look after them … for so long - and [then] they would go off somewhere else. … They were well known to the Health Authority in Dublin … and they were in care and accommodated by the Local Authority in Dublin.
>
> In Dublin … the social worker I met had managed to encourage their participation … educationally. So, one was an avid reader … a Harry Potter fan... Their writing was less practised; [they] were very creative, very capable. … They ran rings round us … the twins [were] identical for us. For example, they would be arrested for each others offences in Britain. … [T]hey were very astute although there were some big gaps in their knowledge. … [I]f there were warrants out … they thought they would be sent straight to prison... They gave us very limited information. Initially it was purely practical resources [we provided] ... chocolate, clothes … buy them a bus ticket.

Here Tarrington corroborates the interview evidence of Jay, Gorman, and Cassidy (cited in Power, 2004) that many Irish Travellers have a very poor understanding of institutional and bureaucratic structures and systems despite their street-wise knowledge. Tarrington (Ibid: 50) continues with his description of the two Irish Traveller children:

> They were twelve, and there was a lot of learning for our Project because they presented as street-wise, as very resourceful. … We had very little understanding of what was actually happening. Where were they staying, were there adults in their lives? They wouldn't say. With hindsight we learned a lot about child protection … because … a twelve-year-old who is sleeping rough in Manchester would cause concern. And yet, they didn't push the same buttons for us as an organisation at the time because of their apparent ease with [the] situation.
>
> So we made a lot of mistakes. … There were major concerns which surfaced later and led to child protection referrals… They as good as admitted they were involved in selling sex. They challenged some of the gay men who worked here [at Safe City] about their sexuality … evidence of their involvement in selling sex, because they were obviously meeting gay men who were approaching them to buy sex. … We said: "Listen we are worried about you because of these particular activities".

> They'd come here and they would get food, a shower, clothes … but again … interaction … was one way and they gave nothing back. … [W]e were saying: "Well we are not a soup run … by all means come and use it, but we actually do more in-depth work - we work around your lifestyle, around your risk…" We built a relationship, but it was managed by them … where they would just run us ragged and turn up at all times of the day. They play[ed] workers off against each other to get what they needed. … They'd bring friends; they'd bring relatives from time to time. We would have a whole crowd of them [Irish Traveller children] just pitch up at three o'clock in the afternoon and all want washing and feeding.

A similar experience was evidenced by staff at a post prison hostel in Manchester (cited in Ibid: 50) who had similar experiences to those of Tarrington with a sixteen-year old resident Irish Traveller girl and members of her immediate family who turned up daily and used the hostel facilities as if they themselves were bona fide hostel residents. Returning to the main illustrative study of two young Traveller boys, Tarrington (cited in Ibid: 50-51) continues his narrative:

> So they presented us with a number of challenges about our own methods of working. Was it fair to say to them: "You can only come and have something to eat if you tell us what you were up to." … They were from the Travelling community… If you asked them a question they would have a discussion in their own language [Cant/Gammon]... We got stuck in a bit of a stalemate with first gate-keeping, and then pushing the boundaries, and us feeling very frustrated. … We were perhaps perpetuating their [risky] existence. … They were selling sex, they were in the city centre at risk of assault … they were [now] fourteen. They were at risk … of solvent abuse, which was massive - absolutely frightening!

The 'Safe City' project eventually decided to make a child protection referral that would have to involve Manchester SSD at the main statutory investigator. Tarrington (cited in ibid: 51) continues:

> They were [then] with another young woman … and we were so concerned about their situation that … we made a child protection referral. … [W]e did it particularly for her, but also for the boys and that changed the relationship utterly - but not in a way that we expected. They had more trust, became more communicative, less chaotic… They felt that … we were trying to do something to change their situation - and we cared. They then learned of my planned trip over to Dublin. We held … a case conference here and invited one of the managers from [Manchester SSD] department to whom we had made the child protection referral. And the social worker from Dublin came over to that case conference. … We didn't get a representative from Birmingham [SSD] - they had closed the

> case … the SSD in Manchester came [mainly due to] the child protection referral...
>
> What we did do was say [to the Traveller children] every time you come to the Project we are going to inform the social worker or call the police. Now in reality that means nothing because the social worker wasn't going to get down here quickly and the police could not be asked. But what we were saying is: "We would take responsibility … we have to tell these people and give them their progress report". [The young woman] was fine with that… We don't do the investigation around child protection - it's [Manchester SSD's] … statutory duty to investigate. They waived their statutory duty to investigate, because they said: "Well, if the young men meet with us then we … will be able to investigate, but until they agree to sit down and do some work with us, we can't investigate it." It didn't progress any further than that and that's quite common. … Manchester SSD weren't prepared to offer services unless the young Travellers would engage on [the SSD's] terms … [and didn't] promise accommodation or even support. What we wanted to do was to set up a package whereby [SSDs in] Birmingham, Manchester, and Dublin would respond.

The logistics of the 'package' would not only mean liaison and co-operation between Manchester and Birmingham SSDs about one hundred miles apart, but also Dublin in another national jurisdiction and on another island. This didn't present a major problem for Dublin SSD who have experience of working with Irish Travellers, but Manchester SSD required the Traveller children to acquiesce to their disciplinary regime before any official engagement could be instituted. The social exclusion of the young Irish Traveller twins and neglect of their pressing needs was effectively the SSD penalty imposed for as Rose (1999: 267) describes '*Those who refuse to become responsible…*'. Tarrington (cited in Power, 2004: 51) continues:

> The [Dublin] social worker involved knew [the Traveller boys] and was their individual social worker… He was prepared to offer services... so they could access a small payment… What [resources] they were getting was from street robbery, stealing from shops, the usual things. So they were at risk in terms of criminal justice activities, [and] due to their lack of resources to keep their health good… One of the [Traveller twins] said to me: "All I want is a flat and a girlfriend, and settle down", an interesting comment… [as] he slept in cars … or warehouses.
>
> They were really chuffed to hear that I was going over to Dublin to meet with their social worker. … They realised we were all working together and we weren't trying to control them - we were trying to provide a support network… They … weren't going to stop moving around, they

> weren't going to stop their lifestyle. "Let's look at how we can make them safe", that was the whole thrust of what we wanted to do. … They weren't old enough to get accommodation on their own and children's homes would have been totally inappropriate for them… The Project leader here [at Safe City] was clear about thinking outside of 'the box'. Let's not apply what we applied to [sedentary] young people indigenous to Manchester. Let's … do something that meets their needs.
>
> When I came back from Dublin, ironically I got a message the next day [that] they are coming back to Birmingham and they have not been back [to Manchester since]. When I was in Dublin I saw a number of extended families who I knew from here [Manchester] - the older brothers and so on - I met a load of them. I also met some of the other young people that they brought with them 'cos they used to just bring someone new each time … [T]hey were really street kids from Dublin and they found it hard work over here because their support networks were [gone]. In Dublin I met them on the street … which was so ironic because to travel to a different country and the young people you work with in this city [Manchester] are there as well. … I visited a project called 'The Loft' which works with young homeless people in Dublin … and had really done very similar things to us. … [They] gradually built a relationship and they were going to be signed up with a package of support for them. They agree to attend 'The Loft' and would get money passed through 'The Loft' - so it's quite a creative way of doing it.

In the above passage Tarrington (Ibid) indicates that these transnational migration routes used by Irish Traveller children between Britain and Ireland are far from unique to these particular children but are used regularly by other Traveller children that have become estranged from their families for various reasons, or whose families have disintegrated for various reasons. These routes may also be used by settled children who have fallen through social security safety nets in Ireland and/or Britain and have developed an association with these young groups of Traveller children. Research interviews (cited in Ibid) with a number of Irish Travellers serving time at Wandsworth prison in London confirm this trend. Two of the offenders interviewed (now in their twenties) spoke about having travelled to Britain from Ireland in their pre-teenage years (when about eleven or twelve years old) and them moving around a number of English cities using informal networks associated with the Travelling communities to network and survive. Again, the most innovative work Tarrington (cited Ibid: 52) encountered was in the Irish voluntary sector, but Dublin SSD had been open to co-operative work with this Manchester based organisation also:

> We hadn't worked with young Travellers before, but we could [have instigated a] child protection referral [earlier but] they don't get investigated properly. … It's a resource led reluctance for Manchester SSD … particularly so in this type of urban inner-city area. I have every sympathy with social workers - you can tell the frustration in their voices. Some of them will justify [this] in terms of [the young person] being stroppy, aggressive, not engaging with the process. ... The investigation would have been them [Manchester SSD] coming to the case conference … That would have been them discharging their duty because they have checked it out. … I don't know how they get away with it sometimes? They were saying that [the young Traveller children] weren't Manchester residents. Now Social Services technically have a duty to any young person.

> [Y]oung Irish Travellers have got left off the list. What we took from it [the Traveller case] was a lot of learning around how we worked. … The biggest thing would be around confidentiality and … that a child protection referral … doesn't mean the end of the relationship - actually it can strengthen it. And we said: "These are our concerns, this is what we are going to do. We want you to be a part of it. If you don't want to be a part of it - that's fine, but we have to do it for these reasons", and [its] usually ok. They saw themselves as Travellers, but even they were marginalised from the Travelling community. … They had been pretty much on their own from the age of ten and had this lifestyle for some time. … Social Services were never going to get there because you needed appointments, a case worker, a social worker assigned to your case work - you had to tick boxes.

The young Travellers that Tarrington (cited in Ibid) struggled to cope with are far from an isolated example. Other Older Traveller respondents (cited in Ibid) revealed that they had similar migratory experiences as children to those discussed above. Two other Irish Traveller men in their twenties described how they emigrated from Ireland to Britain in their very early teens and survived on their wits and with occasional support from other Irish Travellers they happened upon. A thirty-year-old English born Irish Traveller (cited in Ibid: 52) in Wandsworth prison described how he struggled with family breakdown, domestic violence and eventually being taken into care:

> I'm very streetwise, I knew my way around. All the children's homes I've been at I've just ran away from them … and they couldn't control me. They had to put me in secure accommodation … three years of going in and out of those accommodations where I had so much temper. [Due to] domestic violence … I used to run away from home for days and then I'd be arrested for burglary. Most of my crime started because I ran away

> from home and then have nowhere to stay. … I'd have no money for clothes and food so I had to rob to survive and then they caught me again and put me in security in Scotland. ... It stopped me getting back home but ... they take you out on the town. I got out the door and robbed a car. I was about fourteen and drove to London...
>
> When the Social Workers come I'd hide. … I'd get arrested for something and … the Social Services would pick me up. But when I was young I was just too wild with loosing my Gran and domestic violence … it was mad. In ninety-one my mother died and that just led to more crime... I started taking drugs, cocaine, cannabis and drinking alcohol. I was about sixteen ... and she died ... and then all my family started splitting up and things started going wrong. I just carried on committing crime. ... Mainly the drugs was to take away the pain ... it used to cause me inward wrath. [A]s you look back you say [to yourself]: "Well that's not the way to do it".

In Tarrington's illustrative study vulnerable migrant Irish Traveller children were not engaged or supported by two large statutory English Social Service Departments even when their plight was represented to them by a respected charitable organisation [Safe City] and a child protection referral was instituted. The development of Safe City's work with the young Travellers is interesting and instructive about approaches by service providers to constructive engagement with this so-called 'hard-to-reach' group. The development of a level of mutual trust was crucial, as was the setting of 'rules of engagement' between the charity workers and the children. It was particularly interesting how the young Travellers' attitude to the project warmed when the Safe City workers showed real interest in them by instituting child protection procedures combined with open consultation on the issue.

An understanding of Irish Traveller culture by the workers could have led to an earlier resolution of the communication problems initially experienced, but there was no liaison between Safe City and any Manchester based Irish community organisations who work with Travellers. Links need to be developed between organisations working in similar fields to Safe City and specialist Irish Traveller organisations both in Britain and Ireland so that expertise and advice is immediately available when needed. The most disturbing aspect of Tarrington's (cited in Ibid) interview apart from the boys' involvement in drugs and selling sex, and Manchester SSD's inability to engage positively with their needs, were the numbers of other children that Tarrington encountered both in Dublin and Manchester who led similar lifestyles, but who are not protected or cared for due to the institutional blindness and bureaucratic intransigence of some SSDs in England.

Can SSDs in Britain work positively for Travellers?

The Travelling People's Team (TPT) in the London Borough of Haringey is the only established team of statutory Social Workers dedicated to working exclusively with Travellers in Britain. Haringey has had many illegal encampments, so rather than constantly moving Travellers on, the Council decided to develop a rapport with Travellers. In Haringey Gypsy Welfare Officers make first contact with nomadic Travellers in the Borough to ascertain their needs, while anti-Traveller legislation in relation to encampments is not invoked by Haringey Council. Haringey SSD had many reports from the public concerning the neglect of Traveller children, but TPT (cited in Power, 2004) believe that many of these were malicious referrals. The Council is obliged under the Children's Act (1989) to investigate referrals whether malicious or not. TPT now visit people who have made false referrals to inform them about Traveller culture, though most referrals are anonymous. TPT believe that some teachers over-react if they suspect Travellers are abusing their children, and many settled people interpret Traveller ways of living and culture as a threat, therefore malicious referrals are common. Haringey SSD don't actually dealt with many serious Traveller related child protection issues. TPT believe that some Travellers are genuinely interested in the work they are doing, but agreed that Travellers generally dislike social workers. TPT are viewed by Travellers more as Outreach Workers than Social Workers due to their realistic and culturally appropriate attitude to Traveller concerns. TPT do not become directly involved in child protection issues, believing that if they removed a child from a Traveller family this would undermine their work with these communities generally, as TPT have established trusting relationships with Traveller families that would be damaged by acting specifically in a 'traditional' Social Worker role. Therefore TPT's work is mainly supportive of Travellers, though they also deal with domestic violence that sometimes concerns abuse of Traveller elders, and violence between female Travellers.

TPT (cited in Ibid) believe that they meet a specific need for Travellers. Acute social welfare needs are addressed such as housing, disabilities, and domestic violence. When contact is made with a Traveller family, TPT can offer specific supportive help, so that a more holistic approach is encouraged by Haringey SSD. TPT believes that this service is unique as statutory social work teams usually offer crisis intervention only. TPT are trying to build up the knowledge base of local Social Workers around Traveller issues. They work currently with children who have disabilities and special needs. Recently TPT have started development work with young Travellers and women. Community work includes a bi-weekly boys club with some weekend trips away. These boys, aged between twelve and nineteen years, don't attend school, while some are already involved with the criminal justice system. The TPT are setting up a twelve-week course in horticulture for young men and beauty salon training for young Traveller women. A young men's football team co-ordinated by TPT also serves to encourage friendly relations between Irish Travellers and other similar culturally nomadic groups. They play in a league that includes Roma and Gypsy teams. TPT believe that getting groups of boys together on trips and encouraging teamwork through football may lessen inter-community hostility and inter-family feuding, and encourage better inter-Traveller relations generally. A literacy group operates at the Education Centre in Haringey, but TPT believe that Traveller specific literacy classes are needed, though a valid methodology has yet to be developed for this purpose.

Conclusion

The geographical specificity of statutory SSDs in England and their reluctance to offer services to engage with people who don't fit their strict residential or behavioural criteria preclude them from offering supportive services to many Travellers. The managerial, training, morale and resource problems outlined in the Laming Report (2003) that afflict statutory SSDs in many areas mean that marginalised and so-called 'hard-to-reach' groups like Irish Travellers are often omitted from the service agenda.

Irish Travellers are also deeply suspicious of statutory SSDs and perceive them as 'spying' and threatening the break-up of their families. As with Dennis' (cited in Power, 2004) story above, when the school involved SSDs his mother quickly moved location with her family in order to avoid their unwanted attention. The prevalence of these negative perceptions among Travellers are rooted in a long history of cultural insensitivity from SSDs and their association with coercive state assimilation strategies – particularly the removal of children from families.

Tarrington's (cited in Ibid) example reveals a world where Traveller children in their pre and early teenage years lead a homeless and high risk existence at the extreme edge of society due to family breakdowns or estrangements which are usually also characterised by an absence of any sensitive or supportive involvement by SSDs. Tarrington's experience in Dublin and Manchester indicates that quite a few Irish Traveller children have been abandoned by or excluded from society's support systems and migrate between cities in Britain and Ireland living a precarious lifestyle, surviving by sleeping rough, selling sex, engaging in petty crime and whatever else helps them survive, including substance misuse.

It took a voluntary organisation to identify and develop a working relationship with these children, though more speedy progress may have been made if Safe City had engaged the support of a Traveller specific support organisation. Even when called in to effect a child protection order, Manchester SSD failed to act quickly and decisively, instead imposing bureaucratic conditions on any engagement and eventually failing to support the children. Tarrington's (cited in Ibid) engagement with Dublin SSD did result in a positive response and they did have some appropriate strategies in place, something statutory SSDs in England might look to as a development model for working with Travellers. This Dickensian phenomenon of young Irish Traveller children living by their wits in inner city areas with little or no social supports needs to be addressed by the very agencies, SSDs, that in this case eschewed responsibility in at least two large English cities. Manchester SSD is a powerful institution with many resources, but its decision to shun their own duty of care by transferring responsibility for two vulnerable twelve-year-olds to a voluntary organisation can only be construed as a highly inappropriate act on their part. However, Haringey Borough's SSD has developed the TPT that works in a culturally sensitive and innovative way with young Irish Travellers in London.

References

Cemlyn, S (1998) *Policy and Provision by Social Services for Traveller Children and Families*. Bristol: University of Bristol School for Policy Studies: Nuffield Foundation.

Cowan, D and Lomax, D (June 2003) Policing Unauthorized camping, *Journal of Law and Society*, Vol 30, No 2.

Garrett, P (2003) *The Invisible Ethnic Minority: Social Work, Social Services Departments and Irish Children and Families*. London: Action Group for Irish Youth.

Laming, Lord (January 2003) *The Victoria Climbié Inquiry*. Report of an Inquiry, CM 5730. London: HMSO.

Power, C (2003) Irish Travellers: Ethnicity, Racism, and PSRs, *Probation Journal*, Vol. 50, September Issue 03.

—. (2004) *Room to Roam: England's Irish Travellers - a Report of Research*, London: Community Fund.

Rose, N (1999) Powers of Freedom: Reframing Political Thought, in Ruggiero, V., Ryan, M., and Sim, J (eds.) (1995) *Western European Penal Systems*, London: Sage.

Sibley, D (1995) *Geographies of exclusion - Societies and Difference in the West*. London: Routledge.

Traveller Law Research Unit (1997) *Traveller Law Reform: TLAST/TLRU conference and consultation report.* Cardiff: TLRU, Cardiff Law School.

CONTRIBUTORS

Anthony Drummond is a graduate of Social Policy. He is currently a doctoral student at the University of Ulster where he is researching the experiences and the perceptions of Irish Travellers as relating to criminal justice on the island of Ireland, (North and South). In Ulster, he has worked as a voluntary Independent Representative for NIACRO. Prior to coming to Ulster he was employed as a Field Officer for English and Irish Travellers on an action research project funded by the Home Office, delivered by Bedfordshire Rural Communities Charity. Previous to that experience, he was employed as an assistant researcher at St. Mary's University College Twickenham on: *Room to Roam: England's Irish Travellers.* This report can be accessed at:
http://www.irish.org.uk/research_roam.shtml

Professor Ian Hancock is Professor of Romany Studies at the University of Texas at Austin and is a renowned voice in the international Romany movement. He helped petition the United Nations in 1978 for membership of the International Romani Union and in UNICEF in 1987. He is an expert on the Romani Holocaust and served as a member of the U.S. Holocaust Memorial Council under President Clinton. On several occasions, he addressed the Organization for Security and Cooperation in Europe (OSCE) on Roma issues and collaborated with members of Congress on human rights projects. He is the recipient of the Norwegian Rafto Human Rights Prize and the Peace Award of the Bahá'í Church for his efforts to assist the Romani people. He has published a number of articles and books on Roma, including *We Are the Romani People: Ame Sam e Rromane Džene* (The University of Hertfordshire Press,2002), *The Pariah Syndrome: An Account of Gypsy Persecution and Slavery* (Karoma: Ann Arbor, 1987), and *A Handbook of Vlax Romani* (Slavica: Columbus, 1995).

Dr. Mícheál Ó hAodha works as a Lecturer (part-time) at the University of Limerick where he lectures on a number of English, History, Politics and Social Studies courses incorporating Traveller, Roma and Migration Studies. He has published many books about the socio-cultural history, the literature and the postcolonial representation of a number of different (traditionally nomadic) groups within the Irish Traveller community. His publications in this area include: *The Candlelight Painter* (2004); *Parley-Poet and Chanter* (2004); *Canting with Cauley* (2005); *Irish Travellers: Representations and Realities (2006); Road Memories: Aspects of Migrant History (2007)* and *The Nomadic Subject: Postcolonial Identities on the Margins* (2007). He also works in the area of Asylum and Asylum-process - in particular the policy implications and effects of present-day Asylum procedures as applicable within the EU and Ireland – *Postcolonial Identities: Constructing the "New Irish" (2006).* He has been involved in community work with Travellers, asylum-seekers and other marginalised groups both in Liverpool, England and in Limerick, Ireland for the past twelve years.

Ana Oprişan was born in Galaţi / Romania, she lives in Turkey since 2002, she's member of iRSN - International Romani Studies Network and Programme Manager at International Blue Crescent Relief and Development Foundation (IBC-Turkey), working with Roma and other disadvantaged communities in Turkey and coordinating relief operations in Iran and Pakistan. Since November 2005 she has been working on a large post-disaster Relief & Rehabilitation Programme in the Pakistani Kashmir.

Alan Owens was born in Limerick, south-west Ireland. He graduated from the University of Limerick in 2004 with a B.A (honours) in European Studies. Following a period of travel he has returned to the University of Limerick where he is undertaking an M.A. in History.

Dr Colm Power is Senior Postdoctoral Research Fellow in Criminal Justice and Ethnicity in the Centre for Ethnicity and Health, at the University of Central Lancashire. He wrote the Community Fund bid that gained £204,000 for a major social research project on Irish Travellers in England and was Principal Researcher on the project from 2000 to 2004. He is the author of both *Room to Roam: England's Irish Travellers* (Community Fund, London 2004); and *Are You Wide Pavee? An information booklet for Travellers* (Community Fund, London 2004).

David Tuohy is a teacher and a musician – (uilleann piper) who recently graduated from the Irish World Academy of Music and Dance, University of Limerick with a Masters in Ethnomusicology, based on the piping tradition of Traveller musician Johnny Doran.